WITHDRAWN

634.95094
Sav

011539

Warwickshire College

00502757

Plantation Silviculture in Europe

This book is to be ret-

D0322599

011539

Plantation Silviculture in Europe

Peter Savill
University of Oxford, England

Julian Evans
British Forestry Commission

Daniel Auclair
INRA, Montpellier, France

Jan Falck
Swedish University of Agricultural Sciences, Umeå

LIBRARY
WARWICKSHIRE
COLLEGE

OXFORD NEW YORK TOKYO
OXFORD UNIVERSITY PRESS
1997

Oxford University Press, Great Clarendon Street, Oxford OX2 6DP

Oxford New York
Athens Auckland Bangkok Bogata Bombay
Buenos Aires Calcutta Cape Town Dar es Salaam
Delhi Florence Hong Kong Istanbul Karachi
Kuala Lumpur Madras Madrid Melbourne
Mexico City Nairobi Paris Singapore
Taipei Tokyo Toronto Warsaw

and associated companies in
Berlin Ibadan

Oxford is a trade mark of Oxford University Press

Published in the United States
by Oxford University Press, Inc., New York

© Peter Savill, Julian Evans, Daniel Auclair, and Jan Falck, 1997

All rights reserved. No part of this publication may be
reproduced, stored in a retrieval system, or transmitted, in any
form or by any means, without the prior permission in writing of Oxford
University Press. Within the UK, exceptions are allowed in respect of any
fair dealing for the purpose of research or private study, or criticism or
review, as permitted under the Copyright, Designs and Patents Act, 1988, or
in the case of reprographic reproduction in accordance with the terms of
licences issued by the Copyright Licensing Agency. Enquiries concerning
reproduction outside those terms and in other countries should be sent to
the Rights Department, Oxford University Press, at the address above.

This book is sold subject to the condition that it shall not,
by way of trade or otherwise, be lent, re-sold, hired out, or otherwise
circulated without the publisher's prior consent in any form of binding
or cover other than that in which it is published and without a similar
condition including this condition being imposed
on the subsequent purchaser.

A catalogue record for this book is available from the British Library

Library of Congress Cataloging in Publication Data
Plantation silviculture in Europe / Peter Savill ... [et al.].
Includes bibliographical references and index.
1. Tree farms–Europe. I. Savill, Peter S.
SD177.P58 1997   634.9'5'094–dc21   97–15621
ISBN 0 19 854909 1 (Hbk)
0 19 854908 3 (Pbk)

Typeset by EXPO Holdings, Malaysia

Printed in Great Britain by
Bookcraft (Bath) Ltd
Midsomer Norton, Avon

WARWICKSHIRE COLLEGE
LIBRARY

Class No:
634.95094

Acc No:
00502757

Preface

This book is a successor to *Plantation silviculture in temperate regions* that was published in 1986. It is more than simply a new edition. Every chapter has been re-written or heavily modified to bring the subject up to date. Moreover, the welcome addition of co-authors from France and Sweden brings to the text added relevance for Europe and European forestry development. Indeed, this has been a key object in undertaking the rewriting: while still drawing on plantation experience in Britain, we focus strongly on Europe and hope the book is of value across the continent.

Another change also features strongly: plantation silviculture is no longer largely a question of how to grow extensive, industrial plantations successfully, but is also applicable to the many other reasons for establishing trees, including community woodlands, farm forestry, urban forests, energy crops, plantings for amenity and sport, or enhancement of wildlife. We intend that the book applies equally to these more diverse reasons for growing trees, and will help in understanding how to grow them well for whatever purpose.

The focus of our book is on silviculture, but it is a silviculture based on understanding the ecophysiological processes in a forest stand. This underpins the approach and is set alongside the more prescriptive parts. The structure of the book's predecessor was much appreciated and has largely been retained. However, we have enlarged treatment on environmental, biodiversity, and social issues and generally slimmed down the more familiar elements of conventional plantation forestry.

Our aim has been to be concise while providing enough of an introduction for the reader to acquire a thorough overview. References are cited both as source documents and also to allow subjects to be explored further. We hope that the book is welcomed by students at undergraduate, graduate, and diploma levels, and will be valued by managers, and, indeed, anyone with an interest in how best to establish trees by planting and care for the woodland and forest that develops.

Oxford	P.S.
Alton	J.E.
Montpellier	D.A
Umeå	J.F.
May 1997	

Acknowledgements

We gratefully acknowledge the help of colleagues and others who have helped with the revisions of the book, or found the time to comment on individual chapters, and those who have provided literature. In particular, we thank Peter Kanowski who is now Professor of Forestry at the Australian National University, Canberra, to whom we owe most of Chapter 3. It has evolved from research conducted in part for World Bank and IIED reviews, published as Kanowski *et al.* (1992), Kanowski and Savill (1992), and Kanowski (1995). Gary Kerr and Andy Moffat provided valuable advice in revising Chapters 7 and 10; Henri Frochot and Eric Rigolot did the same for Chapters 8 and 13, and Nick Brown, Jeff Burley, and Christopher Quine for Chapters 2, 6, and 12. Michelle Taylor read and helped edit all the text, and Ann-Kathrine Persson and Caroline Benfield typed much of it. The book owes a great deal to generations of students who have discussed many of the topics with us, and whose contributions are considerably greater than they might have realized.

The authors would like to thank the following for permission to reproduce published material:

Institut National de la Recherche Agronomique (INRA) of France and the British Forestry Commission for supplying many of the photographs. These are acknowledged individually under each; Table 14.1 is Crown Copyright and is reproduced with the permission of the Controller of HMSO; Figures 10.8 and 14.3 are also Crown Copyright and are reproduced with permission of the Forestry Commission; the Institute of Chartered Foresters (Tables 5.2, 10.3, Figs 9.1, 9.2, 9.4, 10.2, 10.3, 10.4); the Society of Irish Foresters (Table 7.1);

Table 9.1 is from Cole, D.W. (1986). Nutrient cycling in world forests. In *Forest site and productivity* (ed. S.P. Gessel), page 109, table 6. Martinus Nijhoff, The Netherlands. It is reproduced with kind permission of Kluwer Academic Publishers.

Contents

Part I   Introduction

1 Introduction—the role of plantations

Definitions

Plantations

Plantation forestry has come to be synonymous with high input, intensive management of monocultures for the production of a relatively narrow range of industrial products. In this sense, the development of forest plantation technology has paralleled that of agriculture. The agricultural analogy remains valid where land is not scarce and where the income from this form of plantation forestry exceeds that from alternative land uses. In many other circumstances, particularly where land is scarce, time horizons short, or demand for non-wood products and services is strong, a broader range of plantation objectives and a more intimate integration with other land uses are essential if plantation forestry is to be sustained. This increasing variety of tree planting is seen most notably in the tropics (Evans 1992), but is now also in evidence in temperate regions, including Europe. To promote an understanding of the broader role of plantations, it is necessary to elaborate upon the simple and functional definition of a plantation provided by Ford-Robertson (1971) as 'a forest crop or stand raised deliberately by sowing, planting, or inserting cuttings', to perhaps, as suggested by Kanowski and Savill (1992):

> In its simplest form, plantation forestry describes the intensive management of a forest crop for a limited range of products. There are circumstances in which this definition is appropriate, but many in which complex plantation forestry may be a better alternative. Complex plantation forestry also implies relatively intensive management which controls the origin, establishment and development of the forest crop, but which integrates other land uses within its boundaries, and which promotes the early and continuing production of a wide variety of goods, services and values.

Optimum management strategies for simple plantation forestry are generally well understood, though not necessarily well implemented. Those for complex plantation forestry, designed to maximize social benefits rather than wood production *per se*, are still being developed, and have much to gain from experiences of a wide spectrum of forestry activities, including agroforestry, community forestry, and simple plantation forestry. They are discussed further on p. 229.

The idea of planting a forest, for whatever purpose, may seem simple but unfortunately it is not always clear when a forest stand should or should not be called a plantation. For example when a new forest is established on grassland such afforestation is wholly artificial and may be termed a plantation. By contrast, where an existing forest is regenerated by enrichment, even though many tree seedlings are planted, the general appearance is usually not very like a plantation, at least often not for many years, and would not be called such. A full discussion of definitions is included in the *World symposium on man-made forests and their industrial importance* (FAO 1967), and some of the more arbitrary definitions noted below are derived mostly from this source.

Origins

Between the extremes of afforestation and natural regeneration of indigenous forest there is a range of forest conditions reflecting different levels of intervention in regeneration. Five main types are usually distinguished according to origin.

1. Planting bare land where there has been no forest for at least 50 years. All afforestation of grassland, such as the *Fagus sylvatica* plantations on chalk downland in southern England, *Picea* planting in the uplands (Fig. 1.1), and almost all forest in Ireland fall into this category. The somewhat arbitrary 50 years is used because most vestiges of the previous forest conditions, including the flora, microflora, and fauna will be lost during this period, especially if the land has been subject to arable farming, livestock grazing, or, of course, industrial or urban development.

2. Reforestation of land which has carried forest within the last 50 years but where the previous crop is replaced by an essentially different one. A frequent example in the past was where broadleaved woodland was converted to conifer plantation.

3. Replanting land which has carried forest within the last 50 years by renewal of essentially the same crop as before. This is not as common as the first two because one advantage of planting trees is the opportunity to introduce a new and more productive species. Examples include *Cryptomeria* plantations in Japan, much of the *Quercus* spp. in the New and Alice Holt Forests in Hampshire, England, which has arisen from planting rather than natural regeneration, and most *Pinus* and *Picea* plantations in Scandinavia established in the last 50 years.

4. Forests established by natural regeneration with deliberate silvicultural intervention and assistance, for example many of the *Fagus sylvatica* and *Quercus* forests of northern France and Germany and most conifer forests in Scandinavia and Finland.

5. Forests which have regenerated naturally without any artificial assistance, for example all the remoter conifer forests in north temperate and boreal regions.

Fig. 1.1 Extensive afforestation at Eskdalemuir in south Scotland.

In this book plantations are the forest types in the first three categories above—that is, artificial regeneration is a basic criterion.

Shape

Plantations are mostly of regular shape with fixed and clearly defined boundaries. Land planted with trees is often only called a plantation if its width is at least 100 m, but such a strict definition is not adhered to here.

Stocking

Stocking refers to the number of usable trees per unit area, and the very words forest and plantation imply that land is reasonably well wooded. For a young plantation before it is first thinned a minimum stocking of 1000 trees ha^{-1} or 75 per cent survival of the original planting is generally considered adequate. There are, of course, exceptions to this guideline, such as *Populus* spp. grown for veneer, planted 7 m apart (204 stems ha^{-1}).

Naturalization

New plantations of exotic (introduced) species are obviously artificial and could not occur on the site naturally. But if, subsequently, the species proves well adapted to its new environment, sets seed freely, and can be regenerated naturally (for example *Acer pseudoplatanus* and *Castanea sativa* in Britain, *Pseudotsuga menziesii* in France, *Eucalyptus globulus* in Spain and Portugal), it is said to be naturalized. However, such naturalized species are sometimes considered to be alien by many ecologists, even though they may have been present for centuries.

Mixed regeneration systems

Where enrichment planting supplements existing forest, the forest is normally classified as plantation if the planted trees ultimately form more than half the final crop.

Temperate and boreal regions

This book concerns plantation silviculture in temperate and boreal regions, principally in the cool temperature climates of Europe. This broad climatic zone mostly lies between Mediterranean climates and the cool polar regions. It is characterized by a marked winter period when growth virtually ceases for one to many months because of mean monthly temperatures below 6°C. The period between leaf expansion and leaf fall in deciduous woodlands in lowland England varies between 190 and 240 days, whereas in central Europe it is 140 to 190 days (Jarvis and Leverenz 1983).

Cool temperate climates are in the zone of westerly winds; to be more exact, they are in the path of the depressions which move eastwards from their origins along the polar front (Bucknell 1964). These affect the western coastal areas most and result in a strong maritime influence which leads to equable conditions of warm summers, mild winters, and year-round rain. Away from coastal regions, conditions become progressively more extreme or continental with hot summers, cold winters, and mainly summer rainfall. Plantation silviculture in cool temperate regions is dominated by conifers such as *Picea abies*, *Pinus sylvestris*, and *Larix*, several species from north-western America, and trees from four angiosperm families—Aceraceae, Betulaceae, Fagaceae, and Salicaceae. Some reference is made to warm temperate regions where Monterey pine (*Pinus radiata*) has succeeded so well, along with eucalypts and the Juglandaceae family, with its valuable timbers and often edible fruit.

Silviculture

Silviculture is the art and science of cultivating forest crops for all their diverse range of uses. It is concerned with the growing of trees just as agriculture is concerned with the growing of food crops in fields. Except for some parts of the introductory chapters, this book is primarily concerned with silviculture and not in detail with issues of management, economics, policy, or harvesting.

Uses of forests

Plantations are able to satisfy most of the roles forests play (Fig. 1.2) though they are particularly well suited to providing industrial products.

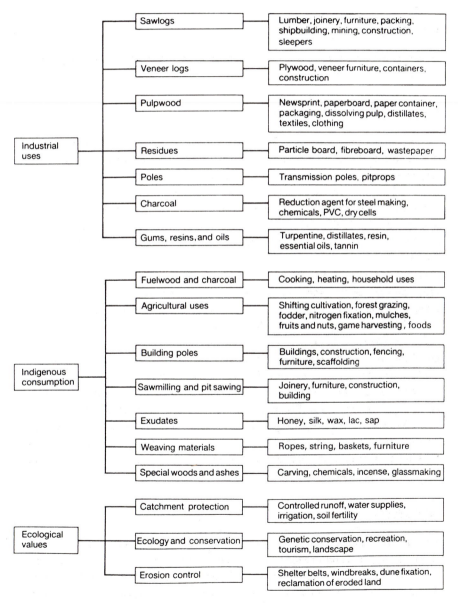

Fig. 1.2   The role of forests (adapted from World Bank 1978).

Brief history of planting

People have been planting trees for a long time; for example many references to tree planting may be found in the Old Testament. In Britain, the most famous early advocate of planting was John Evelyn who urged the repairing of the 'wooden walls' of England in his *Sylva* in 1664 though, in fact, 80 years earlier, concern by Queen Elizabeth I over dwindling supplies of naval timber had led to some *Quercus* planting in Cranbourne Chase. Comparable exhortations and examples can be found from the same period in Germany and France. Colbert, and later Pannelier, was instrumental in the establishment of several thousand hectares of new plantation in the Forêt de Compiegne in the seventeenth and eighteenth centuries. By the early nineteenth century, most texts on forestry included many pages on tree planting and plantation establishment.

Up to the early nineteenth century, nearly all planting was of native species in traditional forest areas; examples in Britain include the New Forest, the Chilterns, and the Forest of Dean, and in France the forest of Compiegne (Fig. 1.3). But increasingly from this time planting involved afforestation of bare land, for example *Pinus pinaster* in the Landes of France and *Picea* and *Pinus* planting on old fields in the Vosges, or conversion of broadleaved woodland to conifer. Many *Larix* plantations and *Fagus sylvatica* and *Pinus* shelterbelts in Scotland date back to this period. But it was in Germany that this new plantation forestry developed most rapidly, largely through the influence of Cotta, such as *Picea abies* afforesta-

Fig. 1.3 One-hundred-year-old planted oak (*Quercus*) and beech (*Fagus sylvatica*) in the Forêt de Compiegne, France.

tion in Saxony. Silviculture in central Europe was, and frequently still is, characterized by very close spacing (planting at least 5000 and sometimes 20 000 trees ha^{-1}),
light thinning, long rotations, and a conservative attitude towards the use of exotics.
Just as today, in the 1800s there were both misgivings about this kind of forestry
(Jones 1965) and enthusiastic supporters such as Simpson (1900) who dubbed this
silviculture 'the new forestry' and urged British foresters to adopt continental practices wholeheartedly.

Concurrent with the development of new ways of growing trees and a multitude
of silvicultural systems to achieve satisfactory regeneration—see all the examples
in Matthews (1989)—was a rigorous evaluation of introduced species as collectors
sent seed and plant material to European countries from all over the world. With
the minor exception of southern beech (*Nothofagus*), all the main exotic species
used in British forestry today (*Picea abies* and *P. sitchensis*, *Pinus nigra* and
P. contorta, *Pseudotsuga menziesii*, and *Larix kaempferi*) were first introduced
more than 150 years ago. In warm temperate regions this trial and testing of tree
species gave rise to the phenomenal success of *Pinus radiata*, particularly in northern Spain and the four southern hemisphere countries of Chile, South Africa,
Australia, and New Zealand.

The present century has seen such an upsurge in tree planting that plantation
forestry is becoming a major forestry activity worldwide. In several countries, such
as the United Kingdom, Ireland, New Zealand, and South Africa, it is the dominant
form of commercial forestry. Historically the first countries to embark on major
plantation programmes were, naturally enough, those seriously deficient in natural
forest cover having reached, as Sutton (1984) describes it, 'the last resort to a
country's wood supply'. Much effort was devoted to species selection, the relatively new challenge of tree planting on bare land, and protecting the expanding
estate.

In Scandinavia, since the 1940s, silviculture has been dominated by planting
(and some natural regeneration) after clear cutting, with rotations of about
100 years. Even in Canada planting is central to its forest renewal programme.
Many factors account for the ascendancy and dependence on plantation forestry,
and these are discussed below. By the end of the twentieth century in temperate
countries there are some 130 Mha of productive plantations which are contributing
an increasingly large proportion of total timber requirements.

Reasons for plantation forestry

Inadequacy of natural forest

All countries have experienced loss of natural forest cover as land has been cleared
for farming and other uses. Although a balance has been achieved between forest
clearance and forest renewal in temperate regions as a whole, in many countries

the loss of natural or semi-natural forest continued long past the point where the remainder was able to satisfy most requirements for forest products. This was the principal reason why several countries embarked on vigorous plantation programmes. However, the total area of natural forest is not the only criterion. Remoteness, inaccessibility, and poor quality of the natural resource may make its working wholly uneconomic and thus indicate other reasons for establishing plantations. In addition, new planting helps to widen the economic base in rural areas and creates employment.

Domestication

This term is used to describe the trend from simply using what forest there is, to growing and managing a specific forest crop that is wanted. The growing of certain *Populus* cultivars derived from a controlled genetic base, for veneer, match-making, or vegetable crates is perhaps the most extreme example.

Plantations as an environmental influence

Considerable areas of tree plantations have been established where the primary objective is not production of wood, though of course this will be an added benefit, but to use the influence which trees and forests can confer. Examples include the role of tree planting in the rehabilitation of industrial waste sites, the shelter provided by windbreaks, stabilization of sand dunes, and, not the least today, as an amenity in urban forestry programmes. Also, today, tree planting to sequester carbon is seen as one of several strategies to mitigate or slow the rise of CO_2 in the atmosphere.

Problems with natural regeneration

On a worldwide scale, natural regeneration, whether planned or not, is by far the most common method of replacing forests. It accounts for about half of the reforestation in the parts of western Europe south of the boreal zone, though less than 1 per cent in the British Isles (Kroth *et al.* 1976), where most establishment is concerned with the afforestation of formerly bare ground. Clearly, opportunities for natural regeneration hardly exist in such circumstances, though it is probable that as first rotations come to an end natural regeneration of *Picea* and other introduced species may become a viable alternative to planting in some parts of Britain.

Frequently, where opportunities for natural regeneration do exist, as in semi-natural woodland, the original species are replaced by planting more productive, commercially desirable trees. Sometimes adequate natural regeneration is cleared and replaced by genetically improved strains of the same species, as occurs with *Picea sitchensis* in Britain (Lee 1990).

Methods of natural regeneration have the reputation of being less expensive than planting. They are especially appealing to owners of woodlands who do not wish to commit themselves to expensive replanting schemes involving clearing and a long period of weeding. Though it may sometimes be cheaper than planting, natural regeneration is not free. Seed years have to be awaited which sometimes necessitates longer than planned rotations. If regeneration is too prolific, respacing must be carried out and if too sparse, gaps must be filled by planting. This unreliability led Sweden, for example, to rely heavily on replanting as the main system of regeneration since 1948.

Where natural regeneration is successful, establishment may be faster than when trees are planted, particularly if advanced growth can be used, and it can benefit the stability of some crops in windy regions (see p. 100). Natural regeneration is also important for the conservation of genetic resources or a particular woodland type.

Although natural regeneration of many species and forest types occurs widely and almost any site is eventually colonized with woody growth, in the absence of excessive tree browsing, it can be difficult to establish a crop naturally at exactly the time it is wanted. This is not a recent problem since planting has been carried out for at least two centuries in many of the lowland broadleaved forests of southern Britain, France, and coniferous forests in Scandinavia (Fig. 1.4). However, as many French, German, and Scandinavian forests demonstrate, natural regeneration is feasible provided there is no urgency to regenerate, and care is taken to ensure favourable stand and ground conditions.

Fig. 1.4 A recently thinned *Picea abies* plantation in Sweden.

Survey of European plantations

The best recent estimates are of between 100 and 135 Mha of simple plantation forests globally, about 75 per cent of which are temperate and 25 per cent in the tropics and subtropics (Kanowski 1995). About 25 per cent of the total is in Europe.

Table 1.1 presents estimates of plantation areas and indicates the rates of afforestation and artificial regeneration for many European countries. In the

Table 1.1 Establishment of forest by artificial means in Europe 1980–1990. Source UN-ECE/FAO (1992), and for Iceland and Norway, Helles and Linddal (1996); •• indicates that figures are not available.

Country	$10^{-3} \times$ forest areas (ha) over 10 years	
	New planting	**Replanting**
Albania	25	••
Austria	25	132
Belgium	••	10
Bulgaria	88	361
Cyprus	2	••
CSFR	58	532
Denmark	10	51
Finland	35	1301
France	455	••
Germany	75	216
Greece	13	••
Hungary	91	95
Iceland	1	••
Ireland	48	20
Italy	91	••
Luxembourg	0.5	4
The Netherlands	12	••
Norway	8	••
Poland	84	615
Portugal	138	92
Romania	2	9
Spain	44	400
Sweden	••	1440
Switzerland	3	19
Turkey	••	••
United Kingdom	246	108
former Yugoslavia	448	••
Total	**2002**	**5405**

10 years 1980 to 1990 over 6 Mha of forest in Europe were artificially established by planting or direct seeding and represented a larger proportion than that created by natural means (4.2 Mha).

Throughout all temperate countries conifers have dominated plantation forestry with the tree genera *Pinus*, *Picea*, and *Larix* being much the most important except in Japan where *Cryptomeria japonica* is the main species. Until recently, the planting of broadleaves has been more localized, aiming to continue a particular forest type or to grow specialized products such as those derived from walnuts, *Populus* spp., and *Salix* spp. Today they are now more widely planted in many European countries as better quality land comes out of agriculture and as concern increases about diminishing supplies of tropical hardwoods. Nevertheless, the main reason for the wide use of conifers is their superior growth rates on the kinds of sites hitherto available for plantations and the need for industrial softwoods.

Plantation silviculture

Plantation life history

Plantations of trees last for a long time and serve many purposes, but all are managed by a similar sequence of operations though not to the same intensity. The main steps in the life history of a plantation are illustrated in Fig. 1.5.

Opportunities and benefits of plantation silviculture

Species

Planting allows foresters a choice of species: they no longer have to be dependent upon frequently random regeneration from existing forest. This is, perhaps, their greatest advantage because the ability to choose species permits the most productive use of sites. At the end of each rotation further improvements can often be made through planting different species, provenances, and genetically improved stock.

Stocking

By planting, a forester can ensure that the whole site is fully stocked for the whole rotation with the kind of tree most wanted. Thus, full use is made of the land available.

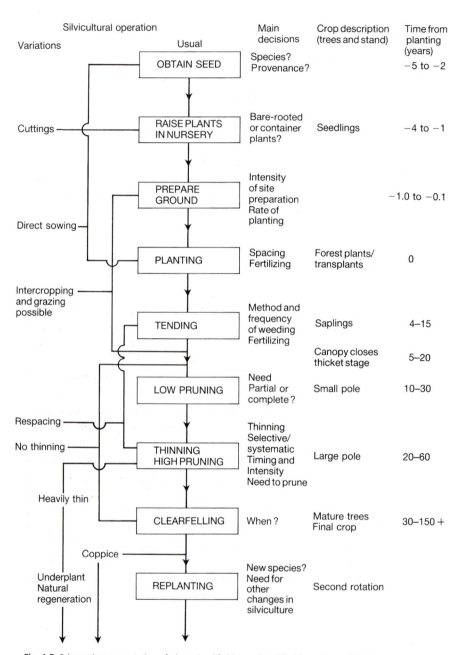

Fig. 1.5 Schematic representation of plantation life history. (Modified from Evans 1982a.)

Product uniformity

Several factors in conventional plantation forestry lead to great uniformity of end product including:

(1) use of one or few species in a stand;
(2) raising a crop to form an even-aged stand;
(3) applying the same silvicultural treatment over a whole stand.

Yields

As a result of these three factors the productivity of plantation forests is almost always much greater than that of natural woodland, see Table 2.1 (p. 22).

Problems with plantations

The intensive nature of plantation forestry demands high levels of commitment, skill, research, and resources. Where these are not available it may be more realistic to rely upon low input and more extensive systems based on natural regeneration, if they can provide the desired species and products. These may be less productive but are more certain of providing a measure of success.

The uniformity of plantation crops brings possible biological risks, instability of trees on exposed sites, and other disadvantages too, particularly aesthetic, and where one woodland type is superseded by another, loss in conservation value. Another concern is that the quality of timber from coniferous plantations may be inferior to that from natural forests and lead to marketing difficulties. This has been particularly true of sawlogs, largely because of the rapid growth rate of conifers. A third concern, on some acid-sensitive soils, has been the role plantations have played in intercepting atmospheric pollutants that lead to increased soil and stream acidity. Some of these disadvantages are discussed later in this book. But, where production is important, they are more than outweighed in most cases by the opportunities plantation silviculture affords for the greater production of wood.

There is no doubt that plantation forestry will become increasingly important. Our understanding of plantations remains imperfect, but their value to us all for countless purposes is unquestioned.

2 Production and long-term productivity in plantations

The main reasons for practising plantation forestry, rather than a system based on natural regeneration, are usually the predictability of yields and the relative ease of control of the production processes. The level of yields and the possible long-term risks involved in achieving them are among the most fundamental considerations in plantation forestry.

Forestry differs from related land uses, like agriculture and horticulture, due to the longer production periods involved. Biomass has to be accumulated, normally over decades, before the trees have any financial value. For most other crops, the time of harvest is highly predictable, sometimes even within days, but in forests it can vary by many years.

The productivity of planted forests varies with region, site, and species but the amount of solar radiation received sets the upper limit for growth. Solar radiation is also used for photosynthesis by potential competitors of the planted trees—shrubs, grasses, and herbs—especially in the younger stages of development. Both competition (see Chapters 8 and 10) and excesses or deficits of water can reduce production (Chapter 5) as can shortages of nutrients (Chapter 9).

Definitions

Primary production is a term used to denote the turnover, expressed as dry weight, in a plant community—the annual production of wood, bark, leaves, flowers, fruits, and roots, whether harvested or not. It may be expressed as:

1. Gross primary production represents total assimilation including the portion subsequently used for respiration in the dark. Gross primary production is closely correlated with the rate of photosynthesis, and hence the amount of light.

2. Net primary production is the amount of assimilated organic matter remaining after the requirements of respiration and other natural losses (e.g. predation and decay) have been met.

Both gross and net primary production may be considered on an annual basis, or over longer periods. The subject has been discussed in detail by Jarvis and Leverenz (1983).

From a commercial point of view the usable portion of this production varies. If the plantation is an energy crop, most of the above-ground net primary production is harvested, but in most circumstances leaves, stumps, roots, and frequently branches are excluded. If only stemwood production is considered, which is normally the case for timber and pulpwood, this commercially usable wood corresponds to about 35 to 40 per cent of net primary production (Cannell 1988). In forestry, production over time is expressed in terms of:

(1) current annual increment (CAI) which is the production or increment in a single year, or

(2) mean annual increment (MAI) which describes the average annual production over a longer period, typically a rotation.

Mean annual increment on any site varies with age (Fig. 10.5, p. 150): it is very low in the years immediately after establishment but it then increases exponentially, reaches a peak (maximum mean annual increment, MMAI) at an age that varies with species and site, and then decreases slowly. The age of MMAI always occurs when the CAI curve crosses the MAI curve.

Primary production is normally expressed in terms of dry weight per unit area, and commercial production of timber as volume per unit area, though weight is becoming more frequently used for pulp and energy crops. The ratio between dry weight (t) and volume (m^3) differs between species from, for example, 0.4 for fast growing *Picea* to 1.2 for *Quercus* in Europe. When heavily branched trees are harvested for energy purposes the larger branches are also included in the volume yield, which is measured in terms of Derbholz—a German term embracing stemwood and branchwood.

The production process

When a plantation is young the stem volumes of the trees are obviously very small, and annual volume production is low because of the small photosynthetic area. Current annual increment depends on photosynthesis during both the previous and current years. At this stage the gap between gross and net primary production is small (see Fig. 2.1), largely because maintenance respiration losses are low.

After a period of growth that is unimpeded by neighbours, crown contact is eventually established between the young trees and the increasing growth rate of individuals slows down. On a unit area basis, rather than for an individual tree, maximum current annual increments are achieved some years later, and are typically 1.5 to 2.0 times the mean annual rates. Fast individual tree growth can be maintained by cleaning or pre-commercial thinning, but excessive growth in the early stages of development may be at the expense of reduced wood quality (see Chapter 10).

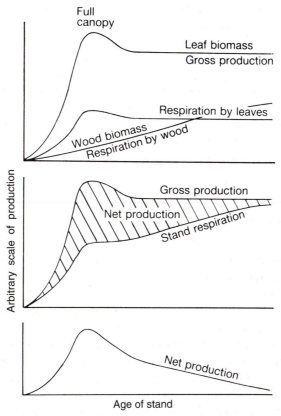

Fig. 2.1 Relationships between gross and net primary production during the development of an even-aged stand. (From Kira and Shidei 1967.)

Ten to 20 years after the first crown contact between trees in many European conditions, the age of maximum current annual increment is reached and annual growth per unit area begins to slow down. During this period, the shaded, lower branches of the trees die and self-thinning begins (see Chapter 10). First thinnings are carried out before competition becomes too severe, usually at a time when the trees have low commercial values: they are usually sold for pulpwood or energy production. Bigger trees (e.g. potential sawtimber) come from later thinnings and these become increasingly valuable, and cheaper to harvest per cubic metre, as their average sizes increase.

If a stand remains unthinned, the gap between gross and net primary production increases by comparison with a thinned stand (see Fig. 10.6, p. 151) due to increased respiration losses and mortality (see Fig. 2.1). However, because thinning can be costly, especially where plantations are remote from wood-using industries, growing trees on relatively short rotations without thinning is quite common and the losses in potential production are accepted.

The time for clear-felling is flexible, often within a range of several decades. The greatest volume, in terms of average production per hectare per year, is achieved if stands are felled at the age of maximum mean annual increment. Something near this age is quite commonly the rotation age for conifers since it more or less coincides with the production of highest value, or greatest net discounted revenue, that is sought by owners (see Chapter 10).

Productivity of plantations

Even the fastest-growing trees transform only a tiny fraction of incoming solar energy into the chemical energy contained in plant dry matter. In most of Europe, production lies within the range of 5 to 20 t ha^{-1} year^{-1} (Cannell 1988). If production at the solar constant of 117.5 MJ m^{-2} d^{-1} could be achieved without reductions caused by, among other factors, the earth's geometry, the proportion of radiation that is photosynthetically active, the quantum efficiency of photosynthesis, the rate of diffusion of CO_2, the proportion of light actually intercepted, and respiration, it would, theoretically, average 26 000 t ha^{-1} year^{-1}.

The highest levels of gross primary production have been recorded in the rain forests of south-east Asia, and are estimated at 110 t ha^{-1} year^{-1} (Kira 1975). However, net primary production seldom exceeds 30 t ha^{-1} year^{-1} in tropical climates because respiration losses are much higher than in temperate regions. On average, over large areas in any natural forest ecosystem, losses from respiration and decay of dead trees or parts of trees balance new growth, so that net primary production is zero: an equivalent amount of carbon to that assimilated in any year is released in the same year. Growth in natural ecosystems therefore gives little indication of the potential of managed forests because the effects of the removal of wood and other products from the forest community are ignored. Net primary production is greatly influenced by removals, by the relative distributions of early and late successional stages, and by other differences in forest structure, especially age. When harvesting begins in mature natural forests, or a windthrow or fire occurs, the balance of ages is changed: young trees colonize openings and grow with few losses, and hence net primary production becomes positive. Generally, leaf biomass and gross primary production rates reach ceiling values at the time the canopy closes but total biomass, and hence community respiration, continue to increase with age (Fig. 2.1). Net primary production therefore reaches a maximum value immediately after the canopy closes, when it often amounts to 25 to 35 t ha^{-1} year^{-1} in dense, young stands with fully closed canopies, and thereafter declines.

On a world-wide basis, in most managed plantations, the average production of stems plus branches is less than 15 t ha^{-1} year^{-1} of wood, and few produce more than 20 t ha^{-1} year^{-1} (Cannell 1988). Maximum yields in north-western Europe are normally in the range of 10 to 12 t ha^{-1} year^{-1}, with the exception of Ireland where 13 to 17 t ha^{-1} year^{-1} have been recorded. In southern Europe, eucalypts can yield

17 to 30 t ha^{-1} year^{-1}. In terms of stemwood volume production, the average for trees in Europe varies widely. For example, *Picea abies* grows at rates of between 3.6 and 14.6 m^3 ha^{-1} year^{-1} in Sweden (Eriksson 1976), while in Britain normal maximum mean annual increments range between 6 and 22 m^3 ha^{-1} year^{-1} (Hamilton and Christie 1971). One of the highest rates of stemwood production known is for *Eucalyptus grandis* at Aracruz, in Brazil, where genetically improved, irrigated, and fertilized clones have mean annual increments of 37 t ha^{-1} year^{-1} (Zobel *et al.* 1983), indicating a level of net primary production of over 70 t ha^{-1} year^{-1} if roots, branches, leaves, etc. are considered as well.

Modelling growth and yield in plantations

Potential commercial productivity over time is often the most important consideration when planning a plantation. Fortunately, even-aged monocultures are well suited to growth and yield studies, and a considerable amount of research has been devoted to this in stands that have reached dominant heights of 9 to 13 m. Some experiments have been followed for very long periods. For example in southern Sweden, *Picea abies* thinning experiments established in 1911 are still being measured on an 8-year cycle, as they have been since they were started (Carbonnier 1954). These sorts of trials provide much of the basic data from which growth and yield models are produced. Such prediction models are indispensable to forest managers for forecasting the effects of possible spacing and thinning regimes, and growth rates on total volume production, and assortments of produce. They also aid decisions about economic rotation lengths (Hägglund 1981).

Such models have been produced for all the commercially important species in Europe (see Chapter 10). For example for *Picea abies* by Eriksson (1976), Møller (1933), Eide and Langsaeter (1941), Braastad (1974), Wiedermann (1937, 1949), Schwappach (1902), Assmann and Franz (1965), Décourt (1971, 1973), and Hamilton and Christie (1971), and for *Pinus sylvestris* by Persson (1992), Andersson (1963), Braastad (1980), Décourt (1965), Vuokila and Väliaho (1980), Wiedermann (1948), and Lembcke *et al.* (1981), and for both species by Agestam (1985), Ekö (1985), Söderberg (1986), and Vannière (1984).

In modern growth and yield models, it is possible to obtain estimates of production (volumes, diameters, and their increments, numbers of trees, dominant heights) at different ages for many initial spacings, site indices, and thinning regimes. In some cases estimates of dry weight are provided (e.g. Eriksson 1976). An example of one such model, for *Picea abies* on a productive site, is shown in Table 2.1.

Evergreen and deciduous forests

Large differences exist between the growth and yield of different species. Deciduous plantations are usually less productive than evergreen ones over long

Table 2.1 Yield Table for Norway spruce (*Picea abies*) in southern Sweden on site index G36 (i.e. dominant height at age 100 years is 36 m) (from Eriksson 1976)

Age (yr)	H_dom (m)	Before thinning d_g (cm)	N (ha⁻¹)	BA (m²)	Vol (m³)	After thinning d_g (cm)	N (ha⁻¹)	BA (m²)	Vol (m³)	Thinning removal d_g (cm)	N (ha⁻¹)	BA (m²)	Vol (m³)	N (%)	V (%)	Total yield BA (m²)	Vol (m³)	MAI BA (m²)	Vol (m³)	CAI BA (m²)	Vol (m³)	Dry weight Mean (t)	Current (t)
20	10.0	9.1	3500	22.6	104	9.1	3500	22.6	104	0.0	0	0.0	0	0	0	22.6	104	1.13	5.2	0.0	0.0	1.50	0.0
26	13.5	11.4	3500	35.5	220	12.9	1807	23.5	146	9.6	1693	12.3	74	48	34	35.8	220	1.38	8.5	2.21	19.3	2.26	4.77
30	15.9	14.5	1807	30.0	217	14.6	1773	29.9	216	7.8	34	0.2	1	2	0	42.4	292	1.41	9.7	1.64	18.0	2.72	5.72
34	18.0	15.9	1773	35.8	288	16.7	1173	25.7	211	14.3	601	9.6	78	34	27	47.9	364	1.41	10.7	1.37	18.0	3.07	5.74
39	20.5	18.7	1173	32.1	298	18.8	1155	32.0	297	11.4	17	0.2	2	1	1	54.3	451	1.39	11.6	1.29	17.5	3.35	5.22
44	22.8	20.3	1155	37.4	383	20.7	884	29.7	305	19.0	272	7.7	78	24	20	59.8	538	1.36	12.2	1.09	17.3	3.61	5.65
50	25.1	22.8	884	36.1	407	22.9	872	35.8	404	15.4	12	0.2	2	1	1	66.1	640	1.32	12.8	1.05	17.0	3.82	5.38
56	27.2	24.6	872	41.1	503	24.7	701	33.7	411	23.9	170	7.6	93	20	18	71.6	739	1.28	13.2	0.91	16.5	4.01	5.57
64	29.6	27.2	701	40.9	541	27.3	692	40.6	537	20.0	10	0.3	4	1	1	78.8	869	1.23	13.6	0.90	16.3	4.17	5.29
74	31.9	29.9	692	48.5	693	30.0	680	48.0	686	22.9	12	0.5	7	2	1	86.7	1025	1.17	13.8	0.79	15.6	4.32	5.27

Total mortality:

Number of stems	Basal area	Volume
355	2.7	27

Note: it is the custom in Sweden to present volumes lost from self-thinning in the 'thinning removal' column of yield tables, hence some of the very low figures for removals of numbers, basal areas, and volumes.

rotations in temperate latitudes, but deciduous leaves are more efficient users of solar energy than evergreen leaves. This apparent contradiction is discussed by Schulze (1982) who pointed out that deciduous leaves have higher photosynthetic rates per unit dry weight of leaf than the more perennial evergreen leaves found on many conifers (10–14 mg of CO_2 g^{-1} h^{-1} compared with 4–6 mg of CO_2 g^{-1} h^{-1}). They are also less expensively constructed in terms of content of assimilated carbon so that a proportionally larger part of the carbon gain by the tree is available for the growth of the non-photosynthetic stems, roots, fruits, etc. Deciduous leaves can be most effective at promoting rapid growth if the growing season is long enough. This fast growth, when young, enables some trees like *Betula*, *Populus tremula*, and *Alnus* to become invasive weeds in young conifer plantations and pioneers in some natural successions, and it accounts for part of their value as energy crops when grown on short rotations (Chapter 15).

In some circumstances evergreen leaves are advantageous because they are retained for long periods, for example often around 5 years and occasionally up to 12 years in *Picea abies*. The evergreen habit enables leaves that are already on the tree to contribute to photosynthetic gain. They can begin photosynthesis as soon as environmental conditions are suitable, even if the new leaves flush relatively late. Because of leaf retention, there is often a much greater leaf biomass, or leaf area index (LAI), in evergreen forests. For example, the LAI in deciduous forests in Japan is between 4 and 7 ha ha^{-1} whereas among evergreen conifers it ranges from 7 to 12 ha ha^{-1} in *Pinus* spp. and 15 to 20 ha ha^{-1} in other species (Tadaki 1966). It has also been suggested that evergreens are at an advantage in environments where nutrients are scarce because the leaves act as storage organs for nutrients in short supply (Moore 1984).

The disadvantage of the evergreen habit is that carbon input, or 'cost' of constructing evergreen leaves, is much higher than for deciduous leaves. It takes several years for a substantial leaf biomass to accumulate. The high structural cost provides protection against heat, cold, desiccation, and possibly predation. It can be justified if the leaves serve for long periods, as is the case for many conifers, but these trees have to accumulate leaf biomass over several years before fast rates of growth are achieved.

Many evergreen trees are more efficient than deciduous ones in temperate climates with a growing season restricted by winter cold and soils with low mineral resources where protection during the vulnerable early years is important (Bryant *et al.* 1983). They can also be very productive in climates with prolonged droughts such as in tropical monsoon and Mediterranean regions. Evergreens predominate in boreal climates but are replaced by deciduous trees where boreal forests give way to tundra. However, in the oceanic arctic region and in continental boreal and alpine climates both forms are successful.

Long-term productivity in plantation forests

Where production of industrial wood is the main objective, the most common plantation forestry strategy involves planting a single, fast-growing, species on a large cleared site, growing it until approximately the age of maximum mean annual increment (Fig. 10.5, p. 150) or to the time of some other measure of economic maturity and then clear-felling the whole crop. This ensures high productivity and is simple to carry out. Felling an even-aged stand at the age when maximum mean annual increment is attained is analogous to harvesting a natural forest at the late pioneer stage of a succession when mean annual net primary production is at its highest. Since monocultures predominate in simple plantation forestry, the species chosen to grow in them are of paramount importance.

Species suitable for monocultures

Although plantations are obviously artificial, the species that have proved success-ful are found commonly in pure even-aged stands in nature. Virtually single-species even-aged forests arise naturally after major disturbances such as fires, windthrow and snow break, floods with associated erosion and siltation, landslips, avalanches, volcanic activity, and insect epidemics. The species that grow in them are, like nearly all of man's crop plants, adapted to colonizing open sites and the large gaps that arise after major disturbances. They are known as light-demanding, pioneer, or shade-intolerant trees and have many of the characteristics of *r*-selected species (p. 165). Practically all the important plantation forestry trees come from this group, including *Pinus, Larix, Pseudotsuga menziesii, Populus, Salix, Betula* (Oliver and Larson 1996), *Picea, Eucalyptus,* and *Alnus.* Thus, the use of such species in plantations is not as artificial as it might appear since natural conditions are quite closely simulated in extensive even-aged, clear-felling systems. For example *Pinus sylvestris* and *Pinus contorta*—two species commonly planted in northern Europe—both become established naturally in pure stands after fires, windthrow, snow break, and severe insect attacks.

In regions where large natural disturbances are rare, there are, of course, very extensive natural forests composed of intimately mixed species and ages. Tropical rainforests provide the most notable examples. Many of the trees that grow in them are adapted to regenerate in closed forests or small gaps. Seedlings survive the much more competitive conditions beneath canopies of more mature trees, and they eventually form understories. These shade-bearing, late successional species are able to persist in conditions of low light intensity, relatively low daytime tem-peratures compared with open ground, and high root competition, especially with subordinate vegetation (Shirley 1945). Late successional trees are characteristically *K*-selected (p. 166) and are adapted to narrow niches in competition with many other species.

There is a continuum between the extremes of light-demanding and shade-bearing trees, and the position occupied by any one particular species may differ with latitude or region. For example *Picea abies* behaves as a colonizing species in large parts of its natural range, though it usually appears later in successions in the harsher boreal climates.

In general, species adapted to regenerate in the crowded conditions under canopies or in small gaps do not make successful pure, even-aged plantation trees, even though they sometimes form pure stands naturally in the older growth stages, by out-competing other species—for example *Fagus sylvatica*, *Tsuga heterophylla*, and *Picea abies* in boreal regions. They can be difficult to establish because they are much more susceptible to frost than colonizing species and also tend to be poorly formed due to reiterative branching. They also tend to grow much more slowly and so require more protection from browsing and more weeding. However, once established they can be very productive. Such species are more effectively managed by silvicultural systems that minimize clear-felling—such as shelterwood and selection systems. They are also valuable for underplanting older crops to obtain forests with two distinct age classes.

Any species, whether colonizing or late successional, that grows naturally in intimate mixture with others is unlikely to make a successful large-scale, pure plantation tree including, for example, *Abies alba* and *Prunus avium*. Many of their defences against attack by pests depend either upon interactions with companion species of plants and animals, or simply by being difficult for pests to find (or 'unapparent'—Feeny 1976), because they are diluted among other species. Trees that grow naturally in mixed forests are relatively safe from serious attacks by predators and parasites so long as they remain growing in an appropriate mixture.

Productivity in the second and later rotations

It is a basic tenet of low-input agriculture that crops should be rotated on the same piece of land to prevent the build up of pests. In forestry changing species at the end of a rotation is seldom feasible, often because suitable alternatives are not available.

A major topic of speculation, debate, and research has been whether the high productivity of first rotation monocultures, especially of coniferous exotics, can be maintained without site degradation, diseases, and serious losses of yields in subsequent rotations. The topic has been the subject of several reviews (e.g. Evans 1976, 1994; Whitehead 1981). It is frequently argued that intensive monoculture is tantamount to challenging the laws of nature (e.g. de Gryse 1955), and so must be doomed to ultimate failure. A contrary point is that trees are traditionally 'soil improvers'.

The debate will, no doubt, continue as new evidence is collected but up to now few significant failures in second rotation monocultures have been recorded in Europe—most of those that have failed are attributed to the incorrect choice of species. However, second rotation plantations (e.g. Fig. 2.2) are still relatively rare, and those with reliable information on growth and yield for more than one rotation, rarer still, so experience is lacking. A few studies have indicated improved growth in second rotations, for example of *Picea abies* in southern Sweden (Eriksson and Johansson 1993). Though observations such as this are common, it is rarely possible to say how much of the improvement can be attributed to better silviculture, improved genetic resources, and possibly a changing climate. More information is available in the tropics. For example in Usutu forest, Swaziland, J. Evans (1996) has assessed the productivity over three rotations of exotic plantations of *Pinus patula*. No evidence of a yield decline as a general consequence of plantation forestry practices was detected there. Over most of the forest the productivity of each successive crop was equal to or better than its predecessor. Only in a small area, where soils are derived from phosphorus-poor gabbro rocks, did some yield decline occur due to phosphorus deficiency. In this case there had been little change in crop genetics or silviculture between rotations, and in the narrow sense of wood production, the intensive plantation silviculture practised appears to have been demonstrably sustainable.

Site and tree stand interactions

When comparing alternatives for regeneration, the choice of species is of paramount importance. In general, broadleaved species are more site demanding than many conifers in that they require nutrients to be available in particular forms (Miller 1984). Though there is relatively little information about the effects of different species on soil processes, some common plantation conifers, especially *Picea* and *Pinus* grown at close spacing with little or no thinning, have a long-

Fig. 2.2 Second rotation, intensively cultivated *Pinus pinaster* near Bordeaux, France. (Photo: J. Gelpe, INRA.)

standing reputation for causing soil acidification and, in susceptible soils, accelerated podzolization. This is accompanied by slow rates of organic matter decomposition and nutrient release and hence immobilization of nutrients in the soil organic matter. It is sometimes claimed that these processes reduce productivity in second and subsequent crops of the same species. A classical and disputed case is the decline in the growth of *Picea abies* in Saxony (Wiedermann 1923). Most examples are confined to sites which are very infertile, or where the species are ill-suited. There is circumstantial evidence of podzolization under *Picea abies* crops in Sweden (Troedsson 1980) and in Britain (Grieve 1978), though in these cooler temperate climates changes in productivity have proved difficult to detect (Holmsgaard *et al.* 1961).

By contrast, many broadleaved trees and a few conifers have reputations as mull-forming 'soil improvers' that can reverse the process of podzolization and improve the rate of litter decomposition. *Betula, Carpinus betulus, Castanea sativa,* and *Larix* are often noted for this. The effects of *Betula* which has colonized heather (*Calluna vulgaris*) moorland have been investigated in Britain by Miles (1981). Heather has a similar reputation to *Picea* for causing podzolization. Miles notes that as *Betula* crops age there is a considerable increase in numbers of earthworms which cause a gradual breakdown of the acid organic material derived from *Calluna*, and convert it to a mull-like humus. This is accompanied by a gradual mixing of the bleached podzolic horizon as the organic material and ferric iron are incorporated, thus tending to a podzolic brown earth. At the same time the rates of organic matter decomposition and nitrogen mineralization increase, as do pH, total phosphorus, and exchangeable calcium in the soil. Earthworms play an important part in mixing soils, and the presence of adequate calcium is essential for their survival (Cooke 1983). In this example, *Betula* has access to sources of calcium lower in the soil profile that are unavailable to heather, because of its deeper rooting and an ability to penetrate ironpans. Because such qualities are associated with many broadleaved trees it is commonly believed that mixtures might benefit the growth of conifers, yet the evidence for any benefit remains largely unconvincing. In most cases the application of fertilizers is likely to be more effective.

Diseases and pests

For a species to survive, grow, and reproduce a set of values of a large number of environmental variables (especially light, water and nutrients) is required and each species responds to it in a particular way. Each species is adapted to exploit one particular set of these variables which ecologists describe as the 'fundamental niche' of the species (Hutchinson 1965; Harper 1977). Niches overlap since most species require many resources. Because resources are limited this overlap results in competition between species and prevents them from reaching their full potential. Thus, niches of newly planted trees commonly overlap with those of aggressive weeds and the trees' survival will depend upon the removal or suppression of the weeds. In natural ecosystems many species of both plants and animals have exerted adaptive pressures on each other over very long periods of time. These pressures involve varying degrees of predation, competition, and symbiosis but usually result in relationships in which none of the associates becomes dominant. Hence relatively stable ecosystems are ensured where fatal outbreaks of insects or diseases are unlikely. Plantations of exotics may be more vulnerable because they have not benefited from interaction with other organisms in the system, nor has a set of parasites and predators of potential pests had time to build up. Possible interactions between species therefore receive a great deal of attention in forestry. They determine the competitiveness of trees with weeds, pests, and pathogens and must not be ignored.

Though the more dramatic losses of plantations are often associated with extreme climatic events, poor technical practice, or poor selection of species (see also Chapter 11), the introduction of diversity into a plantation can sometimes aid the control of herbivorous animals and pathogens. For example *Alnus* spp. have a potential value for the biological control of certain pathogenic fungi such as *Poria* and *Armillaria* in mixed stands. This arises from the ability of *Alnus* to fix atmospheric nitrogen which returns to the litter in nitrate rather than ammonium or amine form, and nitrates inhibit the growth of these fungi (Bollen *et al.* 1967; Li *et al.* 1967). There is evidence from work on *Salix* clones grown for biomass production that plots with mixtures of clones produce significantly more dry matter than monoclonal plots. Figures of 20 per cent or greater increases are quoted for trials in Northern Ireland by Dawson and McCracken (1995) and McCracken and Dawson (1996). This is ascribed, at least in part, to reduced attacks by foliar rusts (*Melampsora* spp.). Disease onset is delayed, build up is reduced, and disease levels at the end of the season are significantly lower than in monoclonal stands. These benefits result from the influence of lower densities of susceptible plants, the barrier effect of disease resistant plants, and induced resistance due to non-virulent pathogen biotypes. However, many of the more widespread and devastating outbreaks of plant diseases are recorded on trees in mixed climax communities (Quimby 1982). They include chestnut blight (*Endothia parasitica*), Dutch elm disease (*Ceratocystis ulmi*), and white pine blister rust (*Cronartium ribicola*) all of which have achieved their status through accidental introductions of exotic pathogens to regions that contain appreciable populations of highly susceptible hosts.

In western interior of Canada, for example, the spruce gall aphid (*Adelges cooleyi*) alternates its life cycle between *Pseudotsuga menziesii* and *Picea*, and mixtures of these two species can lead to infestations of adelgids and reductions in the growth of *Picea* (MacLaren 1983). Another example occurs in mixtures of two-needled pines and aspen (*Populus tremula*), where the rust fungus (*Melampsora pinitorqua*) alternates parts of its life cycle between the two, and can be very damaging to pines.

Difficulties arise in the prediction and identification of future pest and disease problems and these are compounded by the fact that their spread to new regions is still occurring (Gibson *et al.* 1982). Since 1950, one major 'new' pest has appeared about every 5 years on the principal plantation conifers in Britain (Bevan 1984): this may continue for many years, both to indigenous species (e.g. Dutch elm disease) and exotics. Soil-borne diseases caused by wood-rotting fungi such as *Armillaria* spp. and *Heterobasidion annosum* are also quoted as examples that have their origins in forest monoculture systems. In fact, they rely on the presence of infected stumps and root residues and so the time of origin lies more correctly in the history of the site, before planting the current crop, rather than its later use, though because of the short distances between trees of the same species monocultures may influence their development. *Heterobasidion* causes enormous economic losses, but can be controlled to some extent (see Chapter 11).

Though diversity itself is no safeguard against pests, monocultures can favour those with a limited ability to spread, because of the close spacings used in plantations and the readily available food source with little genetic diversity. Risks from pests with wide dispersal ranges are also increased. That the disasters in plantations have not been more serious than the traditional foresters expected may, according to Gibson and Jones (1977), be attributable to two reasons. Firstly, there is usually a wide choice of possible species which allows a good margin for avoidance of the obvious pest and disease problems. Secondly, the economics of intensive systems allow a much greater latitude for expenditure, including research, on protection measures. Thus, the predatory beetle, *Rhizophagus grandis*, was successfully introduced to Britain in 1984 as part of an integrated pest management regime using biological control against the great spruce bark beetle (*Dendroctonus micans*) which had been accidentally introduced in the 1970s. The introduction of *R. grandis* has resulted in reductions of the *D. micans* populations to acceptable levels. Naturally occurring viruses have been used against the pine sawfly (*Neodiprion sertifer*) and the pine beauty moth (*Panolis flammea*), and success is being achieved in the use of insect pathogenic nematodes for the control of *Hylobius abietis*, the major pest of young trees on restocked sites (H. Evans 1996). These are all intermediate pests (see p. 166). It is much more difficult to use biological control measures on pests that have typical 'boom and bust' *r*-strategies. One of these is the northern European spruce beetle (*Ips typographus*) which is periodically responsible for major losses of spruce timber. Beetle populations build up in weakened or recently dead trees associated with windthrow, snow damage, or debris left after felling. If populations are high enough, they attack standing trees invasively and extensive death can occur, especially in large blocks of forest where there are ample breeding sites. Though several natural enemies exist, they are unable to respond to the very rapid increase in bark beetle numbers—which eventually decline more from lack of breeding sites than any actions of their enemies. The most effective way of preventing an outbreak, where practicable, is to reduce the quantity of possible breeding sites by debarking timber left in the forest and ensuring species are well suited to the sites on which they grow.

Exotics

Some parts of Europe lack highly productive native species with timber or growth characteristics suited to plantation forestry, and foresters rely largely upon exotics. Elsewhere foresters have complemented the range of native commercial trees in use with exotics because they can be established easily on certain sites and have better growth rates than native species. The most important exotics used in Europe for timber production include *Pseudotsuga menziesii, Picea sitchensis, Pinus contorta*, species of *Larix, Populus* clones, and a number of *Eucalyptus* species (see Chapter 6).

The reason for the high productivity of many exotics is at least partly explained by the absence of specialized grazing or defoliating insects and leaf diseases, enabling more complete and therefore more efficient canopies to be maintained. Generally the longer a tree species has been present in an area, and the greater the area occupied by it, the greater the number of associated invertebrate species (Kennedy and Southwood 1984), though these relationships are not strong. The relative absence of pests from exotics allows the trees to grow much faster than native species until pests catch up with their hosts, especially if unaccompanied by their natural enemies (see Chapter 11). For example the larch sawfly (*Cephalcia lariciphila*) caused major defoliation in Britain until the specific parasitic wasp, *Olesicampe monticola*, appeared naturally in the outbreak populations in the late 1970s. Within 5 years sawfly populations had been reduced to subeconomic levels (H. Evans 1996). Similarly, a virus disease appeared naturally and controlled the European spruce sawfly (*Gilpinia hercyniae*). Neither sawfly has caused any damage since their natural enemies appeared. If enemies do not appear naturally, a major strategy of control is to introduce them.

The debate about the possible dangers of relying upon exotics is particularly strong in countries with a long history of forest management. In Sweden, Elfving and Norgren (1993) have demonstrated that *Pinus contorta* is likely to grow 32 per cent faster in terms of stemwood volume than native *Pinus sylvestris*, because *Pinus contorta* allocates more of its resources to the growth of stems and fine roots rather than larger roots, as in *P. sylvestris*. This is at a cost in terms of stability. *Pinus contorta*, by virtue of its more plentiful fine roots, is also able to compete better for soil nitrogen, and to use it more efficiently (Norgren 1995). In spite of these apparent growth, and hence economic, benefits fears of eventual pest and disease outbreaks led to legislation in 1979 (Persson 1980) and in 1992 limiting the use of *Pinus contorta* until the possible risks are more clear. However, in Britain, Ireland, and New Zealand, where the native tree flora is either limited, or unproductive, or both, foresters have been less cautious about using exotics in plantations.

Mixed plantations

For the reasons discussed earlier, foresters often try to establish mixed plantations to avoid the investment risks of monocultures. This can be difficult because juvenile growth strategies can differ markedly between species. Frequently, alternating rows of two or more species eventually become monocultures because one species out-competes the other(s). The relatively rare successes arise from timing the planting of each species differently, or using seedlings of different sizes, or being exceptionally diligent in releasing the threatened species from competition by early thinning. In reforestation, mixed species plantations can be achieved more successfully if advanced growth of naturally regenerated seedlings on the site are favoured in pre-commercial thinnings. This strategy is obviously helped if suitable seed-trees are left in the vicinity. Such mixed stands can sometimes be converted to shelterwoods to provide natural regeneration for the next rotation.

Conclusion

To the extent that generalizations are possible or wise, accumulating experience suggests that colonizing species that grow naturally in pure stands are better adapted to monocultures than later successional species. They frequently prove to be successful plantation trees. However, important exceptions exist especially with some late successional species that also form pure stands. These can be useful in more restricted circumstances. The safest strategy is to attempt to simulate what occurs in nature.

In conclusion, it is clear that much has still to be learned about the biological stability and potential problems associated with monocultures. On most sites in temperate regions there appears to be little risk in planting them, but in certain conditions, some of which are known, there is a loss of productivity associated with soil deterioration and attacks by pests and diseases. Constant vigilance is needed to protect valuable monocultures but as knowledge and experience accumulates, so unsuitable conditions will be recognized with greater certainty.

3 Economic, environmental, social, and policy issues in tree planting

While this book is devoted primarily to the scientific and technical aspects of silviculture, they are meaningless in isolation from the economic, environmental, and social frameworks within which decisions about plantation forestry are made. The purpose of this chapter is to describe the latter and to suggest the research and development priorities necessary to adapt current practices to future needs.

Rationales for plantation forestry

The industrial wood production objective of the majority of the world's plantation forests derives from a complex array of forces. Principal amongst these has been the perceived role of forest resources in economic development—eloquently enunciated by Westoby (1962), and subsequently reviewed by, amongst others, Douglas (1983), Westoby himself (1987), Byron and Waugh (1988), Binkley and Vincent (1992), Kanowski and Savill (1992), and Evans (1992) devoted a chapter to 'Why plantations in tropical countries'. Economic development theory provides a rationale for the establishment of plantation forests for both industrial and non-industrial purposes. Plantations may provide:

(1) a resource base for the development of wood-using industries. Plantations may represent a new resource (e.g. Portugal) or a replacement for diminishing (e.g. United Kingdom) or less productive (e.g. Australia, Sweden, Finland) natural forests. The economic rationale may be one of import substitution (e.g. Australia, United Kingdom) or generation of export income (e.g. New Zealand, Swaziland);

(2) an energy source, for either industrial (e.g. wood-gas—Sweden, Finland, Ireland) or non-industrial (e.g. fuelwood, charcoal) production (Christersson *et al.* 1993; Davidson 1987);

(3) rural employment in circumstances where labour is available and its cost is relatively low. In economically developed countries, the historically high levels of employment required for plantation establishment and management

have been supplanted by mechanization (e.g. Stewart 1987). However, in economies where labour costs are relatively low, plantation programmes continue to have considerable potential to generate employment (Davidson 1987);

(4) maintenance or enhancement of agricultural productivity (e.g. Rosenberg 1974; Felker 1981).

In other cases, environmental concerns have prompted plantation establishment for:

(1) environmental protection or rehabilitation, especially soil and slope stabilization (e.g. Switzerland);

(2) recreation and landscape values (e.g. The Netherlands, Belgium) and also for promoting public health (e.g. Sweden);

(3) protection of natural forests from exploitation for either commercial use (e.g. Australia—Cameron and Penna 1988) or subsistence needs;

(4) carbon sequestration in an attempt to mitigate global warming.

Governments have often sought to promote plantation forestry for one or more of these reasons. They have done so directly, by financing plantation afforestation programmes through, state forestry agencies (e.g. Australia, China, New Zealand) or parastatal, and indirectly, through grants, subsidies, or tax concessions (e.g. United Kingdom). Where large private investors are involved in plantation forestry, they are likely to be motivated by the expectation of attractive financial returns, over the longterm, relative to other investments.

Although plantations have usually been developed on land owned or managed by the state or large-scale forest industries, smaller private landholders have been regionally important (e.g. New Zealand—McGaughey and Gregersen 1988), and are likely to become more so in the future as pressures for land increase. There is a wealth of information describing the characteristics and motivation of these forest growers in both industrial and non-industrial societies (e.g. Arnold 1983; Blatner and Greene 1989; FAO 1985; Grayson 1993; McGaughey and Gregersen 1988). Their rationales for plantation establishment may reflect, to varying degrees, elements of those outlined above. They may also include:

(1) productive use of land which has become surplus as a consequence of altered farming systems or reduced labour availability (e.g. Hummel 1991): in some cases, plantation establishment may reinforce individuals' land tenure rights (Gondard 1988);

(2) income generation opportunities in response to local market shortages, however transitory (e.g. India—Saxena 1991);

(3) investment strategies which emphasize capital accumulation, in the tree crop, rather than income generation (Arnold 1983; Chambers and Leach 1989).

In reality, these rationales often interact and any individual plantation is likely to be established as the result of a variety of these forces.

The policy context and its elements

Decisions about how forests and trees are used are largely determined by pro-claimed or *de facto* policies which govern land use, land tenure, forestry activities, as well as forest industries and markets, and by policies bearing on those activities which transform forests to other uses. The circumstances most favourable to the successful development of plantation forestry are those in which a co-ordinated land use strategy has been formulated and agreed, tenure rights are unambiguous and unchallenged, forestry is treated as more than a residual land use, the relative roles of natural and plantation forests are well defined, and forest industries or markets are sufficiently developed and stable to maintain a relatively assured demand for forest products. The financial and biological characteristics of trees as plantation crops also have important consequences for the formulation and imple-mentation of policies relevant to plantation forestry.

Trees as plantation crops

The primary distinguishing feature of trees as plantation crops, compared with agriculture and horticulture, is their relatively long production period. In temperate countries, this is seldom less than 8 years, rarely as short as 20 years, and tradition-ally at least 50 to 100 years or even longer for prime hardwoods such as *Quercus* and *Fagus sylvatica*. Successful plantation programmes, whether on a large or small scale, therefore require the commitment of adequate resources over a consid-erable time period, and depend on security of land tenure, assured rights to the tree crop, strong and continuing technical input at the appropriate level, and linkage to markets.

Financial characteristics and consequences

In the case of most simple forest plantation crops (see Chapter 1) grown primarily for wood production, costs—planning, acquisition and raising of planting stock, site evaluation and preparation, planting and maintenance—are concentrated in the early phases and returns from harvesting in the late stages of the production period. There are annual costs too, for taxes, insurance, forest protection, and management. However, as financial evaluations based on present value criteria are strongly influenced by the time between investment and return, this period—approximated for most plantation species by their rotation length—is therefore critical in determin-ing the financial returns from tree growing. The major consequence is that private investment in plantation forestry can probably only be expected in economies which are perceived to be stable over the period of investment (Yoho 1985), and for species and sites from which favourable financial returns are most assured.

Governments have, therefore, often been seen to have a leading role in undertak-ing or facilitating afforestation by planting. However, financial returns from planta-tions are not necessarily poor: McGaughey and Gregersen (1988) suggested that the return on investment for well-managed, fast-growing, industrial-tree crops should be in the order of 10 to 15 per cent, consistent with the 8 to 12 per cent internal rates of return reported by Elliott *et al.* (1989) and Whiteside (1989) for New Zealand *Eucalyptus* and *Pinus radiata* plantations. Although returns from slower-growing plantations in cool temperate regions are probably more in the order of 2 to 5 per cent (e.g. Wilson 1989; Spilsbury 1990), they nevertheless compare favourably with real rates of return from other long-term investments (Leslie 1987). The returns from any particular plantation programme will depend on a myriad of biological (e.g. growth rate) and economic (e.g. capital and labour costs) factors, as well as associated environmental (e.g. conservation, landscape, hydrology) and social costs and benefits.

Although the issue of public or private ownership has been hotly debated in many countries (e.g. New Zealand—Kirkland 1989; United Kingdom—Rickman 1991), it is probably more a matter of politics than economics. The successful involvement of both large and small-scale private investors in plantation forestry in many countries (e.g. United Kingdom—RICS 1996) suggests that factors other than the economic characteristics of trees themselves are the primary determinants of the level and source of investment; a secure investment environment and sufficient investment capital are probably most critical. Depending on the scale of private firms' or individuals' holdings, taxation arrangements which do not discriminate against forestry or the availability of credit on appropriate terms for tree growing may be important in facilitating plantation forestry (Arnold 1983; McGaughey and Gregersen 1985; Brunton 1987).

The foregoing discussion assumes that investment decisions are made primarily on the basis of discounted net benefit, which is likely to apply to many large investors. There is considerable debate over the more general relevance and application of this criterion (e.g. Leslie 1987; Price 1993). In any case, for many small-scale growers, the risks and cash flows associated with plantation forestry are more likely to determine their behaviour than are evaluations based on net income over a long time period. For owners of small areas of plantations, a well balanced distribution of age classes that assures a steady income can be very important. Schemes by which industrial wood users enter into agreements with private landholders for the supply of wood are often successful (e.g. Portugal), and may involve the wood purchaser in establishment and management of the plantations. In such cases, there is usually the added advantage of more efficient and more satisfactory operations, in both environmental and social terms.

Biological characteristics and consequences

In circumstances where complex plantation forestry is appropriate, the biological characteristics of trees may be more important in determining plantation success than their financial characteristics. Species and management regimes which produce returns only after a long period, or which do not integrate well with other land uses, may be inappropriate. The latter may include, for example, collecting berries and fungi, values for hunting, or a pleasant environment for hiking. In some cases, the adoption of complex plantation forestry may require only a modification of traditional silvicultural principles (e.g. Gilmour *et al.* 1990). In others, a synthesis of agroforestry and plantation forestry approaches will be desirable (e.g. Sargent 1990); elsewhere, replacement of species which produce only wood by those which also yield valuable intermediate returns may offer the best solution (e.g. Keresztesi 1983). Depending on the tree species and silvicultural regime, a variety of agroforestry practices (e.g. intercropping, underplanting, or rotational cropping) may be feasible, according to particular circumstances and demands. The relative merits of each will depend on the biological and economic characteristics and interactions of the tree and non-tree crops, the sustainable land use systems with which local people are familiar or which can be introduced, and the risks and cash flows associated with each option. The use of 'multipurpose' species may be dependent upon developments in technology to facilitate multiple uses.

Land use and tenure contexts

Land use policy, and its explicit or implicit expression of the role of forests and forestry, forms the basis for forest planning and management. Security of tenure and of rights to forest products, essential for successful plantation production, do not imply that the forest or its products must necessarily remain in one ownership over the production period, but that the fundamental elements of property rights necessary for efficient market-mediated transfer—exclusivity, transferability, divisibility, and enforceability—should apply (Binkley and Vincent 1992; Randall 1987).

Policy formulation and plantation planning

Whatever the policy context, it is widely acknowledged (e.g. Barnes and Olivares 1988; Douglas 1986; Winpenny 1991) that the successful formulation and design of forestry interventions demand more time, effort, and flexibility than have commonly been allowed in the past. In many cases, the ultimate success of plantation forestry proposals will depend on policy makers' and planners' acceptance of these requirements.

Implementing plantation forestry

Plantation forestry should be implemented sustainably in ways that encompass its biological, economic, and social dimensions (e.g. Ascher and Healy 1990; Barbier 1987). Although there are many difficulties in the definition and evaluation of sustainability (e.g. Cocklin 1989; Caldwell 1990; Upton and Bass 1995), the concept describes a framework within which acceptable plantation practice can be defined.

Within this framework, effective planning and management are critical to the success of any plantation programme. Both involve iterative processes of defining objectives, formulating alternative means of attaining them, estimating their impacts and assessing their costs and benefits, revising proposals, and selecting, implementing, monitoring, and redefining, the best option(s). In the sections that follow, the co-ordinated sequence of activities necessary for the successful conduct of plantation forestry are outlined.

Planning, impact assessment, and monitoring

Given a favourable policy environment, and assured tenure or rights to trees and/or land, the principal requirement for the development of successful plantation forestry is reliable information describing the land resource and its biophysical characteristics, current users and use patterns, and relevant social and economic issues. The form in which this information is presented is important; it must be accessible to the range of interested parties (Davis-Case 1989). Much of the geographic and biophysical data can be usefully presented as maps, at a scale of 1:10 000; the development of geographic information systems has greatly facilitated their processing and presentation. The form in which other information is presented will depend on its nature and intended audience.

These data should form the basis for preparation of draft proposals for review, usually by appropriate state agencies, and at least those individuals or groups directly affected by, or with a recognized interest in, the planned development. A planning process which invites comment at this stage, and is seen to respond to it, is more likely to gain eventual acceptance than one which does not. Numerous ideotypes and procedures (e.g. Cassells and Valentine 1990; Davis-Case 1989; Knopp and Caldbeck 1990) have been suggested or implemented to address this challenge.

Plantation planning

Successful plantation establishment and sustainable management demands integrated planning at all levels. Minimum planning requirements for plantation forests are (adapted from ITTO 1991):

(1) clear definition of objectives—plantation forestry involves the management of trees and forests to achieve certain defined human objectives; it is only after they have been specified that appropriate sites, species, and technologies can be proposed and evaluated;

(2) recognition of legal requirements, customary rights and local demands, and their consequences for forest composition, design, and management;

(3) prior survey to identify and protect areas of conservation and cultural significance, physical fragility, landscape value, and hydrology—these areas should be reserved from afforestation and protected from damage;

(4) plantation design to retain a network of natural vegetation, be in sympathy with land forms, and avoid damage to watercourses—the location and construction of access and extraction roads and of fire protection networks is particularly important in this context;

(5) definition of site preparation, establishment, management, and harvesting regimes which minimize adverse environmental impacts—this can require co-operation between adjoining plantation owners, if only to minimize the risks of windthrow and insect epidemics (Persson 1975; Burschel and Huss 1987);

(6) definition of plans for protection against and management of fire, biological pests, and diseases;

(7) derivation of product utilization and marketing strategies.

Assessment of impacts

Plantations are established within a physical matrix which may be loosely termed the 'site', which is a complex of physical, chemical, and biotic factors. Physical factors include the soils in which plantations are established. They can vary according to geology, topography, and elevation. Chemical factors include soil fertility and atmospheric pollution. Biotic factors include the plant and animal communities which the plantation might influence, or which might affect the growth of the trees, and the human communities around which plantations are established. Conversely, plantations influence soil development, microclimate, surface water pH, and carbon storage, among other things.

Whatever the objectives and management strategy adopted, the maintenance of environmental quality and integration into local societies and land use practices are minimum requirements for plantation forests. Environmental impacts are typically identified and evaluated through preparation of an environmental impact statement (Hyman and Siftel 1988), and usually addressed by the definition of operating standards and codes of practice (e.g. Forestry Commission of Tasmania 1989; ITTO 1990, 1991; Poore and Sayer 1987). Establishing the plantation programme in the broader social context is somewhat more challenging, although appraisal and diagnosis methodologies (e.g. Formby 1986; Raintree 1987; Molnar 1989) have been developed to offer some guidance. The sensible and sensitive use of appropriate impact assessment methodologies should form the basis for evaluating plantation proposals prior to their adoption.

Auditing and monitoring

Environmental auditing and a long-term monitoring system are both necessary to describe plantation impacts objectively, preferably quantitatively, and to assess the sustainability of plantation crops over successive rotations. In addition, an objective means of monitoring not only provides a way of detecting long-term changes, but it also has many short-term managerial uses, such as the preparation of yield models and estimating future production. It can provide a means of detecting nutrient deficiencies, pest outbreaks, and other problems which might be unknown or unsuspected at the time of establishment.

The design and conduct of auditing and monitoring systems have been described elsewhere (e.g. Shell/WWF 1993). Whilst initiation of a monitoring programme is usually the task of a specialist, its maintenance is neither expensive nor difficult. It is essential that field managers appreciate the reasons for its existence and its potential value, and are therefore motivated to maintain it. Since the mid 1980s there has been an increasing interest and debate about the desirability of participating in voluntary external and independent auditing of forests against standards previously agreed as significant and acceptable to stakeholders. The market-based objective of having forests 'certified' in this way is to improve market acceptability and share for the products of good management (Upton and Bass 1995).

Elements of plantation technology and management

Contemporary forest management has a certain dualism: one element is the sophisticated, quantitative management science which has been developed for large-scale simple plantation forestry. Planning and analysis methodologies are usually based on mathematical optimization, and have been developed for purposes such as multiple-use planning, harvest scheduling, and evaluating silvicultural alternatives. Approaches and systems are well described by, e.g., Clutter *et al.* (1983), Dykstra (1984), Leuschner (1984), and Whyte (1988). However, the simplifications inherent in mathematical modelling are also a limitation of many of these approaches, which have been criticized for their emphasis on wood production (e.g. Behan 1990; Maser 1990).

In parallel, there has been increasing acknowledgement of the value of traditional management systems as similarly sophisticated, albeit qualitative, and the development of management practices which build from them. The progressive understanding of these systems and of their interactions with societies are recorded in the development literature (e.g. Arnold and Stewart 1991; Arnold 1991; Cook and Grut 1989; FAO 1985). This evolving understanding has much to contribute to the development of complex plantation forestry.

The technical elements of plantation forestry apply regardless of the simplicity or complexity of the activities, and are shown in Fig. 1.5 (p. 14).

The implementation of these activities follows from, and takes place in, the context of the planning, assessment, and monitoring processes described above. Although each can be represented, discussed, and implemented independently, it is their integration into a co-ordinated system which will make plantation forestry successful. Foresters are continually having to choose between different silvicultural and management systems to achieve different mixes of products and benefits from specific forest areas in particular circumstances: there is no single system ideal for all situations. Where the necessary resources and management expertise are available, the productive capacity of plantations is unlikely to be rivalled; where such interventions are less certain, so should the expectations of and reliance on plantations diminish.

Successful plantation forestry

Successful plantation forestry depends, therefore, on the co-ordinated application of an array of technical elements, integrated within the broader economic, social, and cultural contexts. Each plantation is to some extent unique and site specific. Whilst the fundamental principles of successful plantation forestry are well established, their successful application to particular situations requires informed interpretation and intelligent adaptation. Successful plantation forestry is therefore characterized by substantial and sustained research and development efforts specific to the particular plantation programme. It is clear to us, however—as it was to Byron and Waugh (1988) in the case of forest resources more generally—that issues of political economy are far more important determinants of the nature and success of plantation programmes than are issues which are primarily technical.

Unfortunately, assessments of development projects typically focus on financial and technical impacts rather than on their broader consequences in environmental, institutional, and social terms (Barnes and Olivares 1988). Analyses of plantation forestry proposals or practices should, therefore, first address these issues; it is only then that appropriate scientific and technical interventions can be determined.

Improving plantation forestry

The earlier parts of this chapter have outlined the nature and contexts of plantation forestry. The remainder of it discusses how plantation forestry might be improved, in terms of both its role within societies and economies and in its technical practice.

Simple plantation forestry

'Plantation forestry' has come to imply relatively intensive management of simplified forest systems for a range of wood products. In terms of wood increment per unit area, well-managed, simple plantations are many times more productive than most natural forest systems, and the plantation forests of many countries are now a major component of their productive forest resources; indeed, plantation products are expected to take an ever-increasing market share. The financial returns from these plantations can compare favourably to those from alternative, sustainable, land uses.

Such plantation forestry is relatively demanding of resources. Consequently, it is increasingly concentrated on those sites which are inherently most productive, and where environmental and social values are least likely to be prejudiced. In effect, this implies less extensive, more intensive plantation programmes, building on an enormous potential for genetic improvement concentrated in countries where the forest land base is stable and secure (Bingham 1985; Gauthier 1991).

Complex plantation forestry

In many other circumstances, plantation forestry will have a broader and more integrated role. This more complex form of plantation forestry will be distinguished by objectives, species, management regimes, and tenure arrangements other than those typical of simple plantation forestry. Its major characteristics are likely to include:

1. A more intimate association with other land uses—simple plantation forestry is typified by a sharp distinction between plantation forest and other land use. The 'boundaries' between plantations and other uses will become less distinct as plantation forestry becomes more complex.

2. More direct involvement of local people in the conception and implementation of plantation forestry and in the sharing of its benefits and products—simple plantation forestry has been characterized by the limited involvement, for example as wage labour, or exclusion of local people (e.g. Douglas 1983). There is an increasing understanding of how participatory planning, management, and use might be developed and practised (e.g. Arnold and Stewart 1991; FAO 1985), and this approach now characterizes some programmes involving plantation forestry (e.g. Gilmour *et al.* 1989). In some cases, altered tenure arrangements may facilitate complex plantation forestry (e.g. Arnold and Stewart 1991; Sargent 1990).

3. More complex species composition and plantation structure—simple plantation forests are typically even-aged monocultures, producing similar products discontinuously. Complex plantation forests are likely to comprise more intimate mixtures of species, yielding a more continuous flow of more diverse products. This does not necessarily imply that tree species will grow as polycultures (e.g. agroforestry), though this will be appropriate in particular circumstances. In others, a mosaic of relatively small blocks of tree species may be more easily managed, but still yield the desired variety of products.

The greatest challenge to the adoption of complex plantation forestry is probably the necessary conceptual change, or 'paradigm shift' (Gilmour *et al.* 1989), on the part of policy makers and forest managers. There is, however, encouraging evidence of rapidly increasing awareness of the potential of complex plantation forestry in diverse circumstances, and of its implementation. For example Spanish

Eucalyptus plantations are now being designed and managed to facilitate a range of agroforestry practices and yield a variety of products—e.g. sawn timber, honey, and eucalyptus oil—in addition to pulpwood. In Britain, the new National Forest and Community Forests (Countryside and Forestry Commissions 1991) are based around plantation forests managed for wood production, but with major emphasis on design and management for recreation, amenity, and conservation values, as are many privately-owned forests (RICS 1996). Plantations are also used as 'offsets' for carbon storage, to compensate for western energy consumption as well as a means of producing a renewable form of energy in the form of short-rotation coppice (Chapter 15), and in temperate countries planting rates are increasing partly to compensate for tropical deforestation.

Conclusions

The appropriate role and future prospects of plantation forestry vary between and within nations. Although there are prospects for global and regional co-ordination of forestry activities, decisions about particular plantation forestry programmes can only sensibly be made at levels close to the resource users. Strategic reviews of plantation programmes and prospects are common at the national level (e.g. Dargavel 1990), and provide a means of linking international opportunities and concerns with local experiences, possibilities, and constraints. The conclusions drawn below therefore provide only a framework for policy determination, scientific research, and operational implementation:

1. Plantation forestry is most appropriate where land and tree tenure are uncontentious, and where plantation systems can be managed sustainably to satisfy both industrial and non-industrial demands. Simple plantation forestry, emphasizing industrial wood production, will become increasingly concentrated on productive sites close to processing facilities, in nations whose economies are perceived to be relatively stable.

2. In many other circumstances, investment in complex plantation forestry is more likely to satisfy societies' demands from forest resources than is investment in simpler systems. This implies a redefinition of the objectives and principles of plantation forestry to encompass more integrated forms of land use yielding a wider range of products.

3. Proposals for plantation forestry should therefore be formulated in the broader social context. Participatory methodologies developed for agroforestry offer some guidance to this. The assessment of social and environmental impacts, and their acceptance by those affected, are prerequisites for successful plantation forestry.

4. Decisions about investments in plantation forestry should be guided by broad economic assessment of plantation performance and alternatives, rather than

by the narrow financial analyses often applied in the past. The evolving economic appreciation of the elements of sustainable development has much to
contribute towards this end.

5. Conceptualization, planning, and implementation of plantation forestry pro-
    grammes must emphasize flexibility and adaptability, which should also char-
    acterize the terms of finance for plantation forestry.

6. Integrated, interdisciplinary research is fundamental to successful plantation
    forestry, both prior to plantation design and establishment and throughout the
    plantation life cycle. Such research would define the economic, social, and
    environmental contexts of possible plantation forests, underpin their silvicul-
    tural excellence, and draw from the substantial body of scientific and techni-
    cal information already acquired in most regions of the world. The integration
    of agroforestry and plantation forestry methodologies and technologies may be
    particularly valuable.

7. Well-qualified, well-motivated, competent, and imaginative operational staff,
    sensitive to local demands and expectations, are essential for the implementa-
    tion, monitoring, and adaptation of plantation forestry, particularly as it
    becomes more complex. Their education and training should emphasize skills
    in the recognition and analysis of issues and problems, including social and
    economic ones, as well as the scientific understanding and technical com-
    petence necessary to raise, establish, maintain, and monitor plantation stands.

The relatively short history of plantation forestry has been characterized by a
focus on the scientific and technical requirements for wood production, which have
been successfully developed and implemented in many cases. Often, however, the
economic, environmental, and social consequences of plantation forestry programmes have been insufficiently emphasized. The potential contribution of plantation forestry to most societies will only be realized if its purposes are broadened,
and its practice more integrated with other land uses. The challenge ahead is, therefore, to build on the considerable body of experience and information relevant to
plantation forestry, and direct it towards a broader range of objectives and practices
than has been common in the past.

4 The layout and design of plantation forests

In laying out plantations consideration must be given not only to the well being and economic management of the plantations themselves, but also to wider land-use issues. This is particularly important for large plantation developments. Forests are no longer seen as places merely having physical, biological, and economic potential for growing wood, that need design features to protect them from fires and wind, and to provide access. Today, the plantation manager is more of a tenant of the land and a custodian of its other values and possible uses as well. This awareness emphasizes the obligation for wise use of the land set aside for forest plantations. Forests have visual and aesthetic values; they are habitats for many species of wild plants and animals. They have other economic and social functions, such as recreation, water gathering, hunting, and some forms of agriculture. Forest may, too, contain important archaeological features or be an integral part of a much prized landscape. These values can conflict with growing wood as an economic crop and, in achieving a balance, compromises must be made.

One of the principal influences affecting these other values is the arrangement of plantations in terms of which areas should be planted and which should not. A secondary consideration is how species and ages might be varied to accommodate good landscaping and conservation principles, to facilitate the management of wild animals, to enhance woodland amenity, and so on. In plantation design, both for new planting schemes and modifications to existing ones during the felling phase (restructuring), surveys are usually carried out so that decisions on these matters have a rational basis.

There is no standard recipe for designing a forest since every case is unique. What may be appropriate in one place, or country, may well not be in another as economics, site, topography, and other conditions change. However, in many countries there are published guidelines, for example Forestry Commission (1994, 1995) and Office National des Forêts (1989), that provide advice or lay down minimum standards of practice particularly for sensitive issues of conservation, landscape, and the environment. The main factors to consider are discussed below.

Where to avoid planting

Much of the analysis in this chapter recommends practices that reconcile good silviculture with other benefits from a forest. However, the first consideration is to determine where plantation establishment should not be attempted because of a pre-existing value such as a special conservation site or archaeological remains. Full and early consultation with interested parties is essential; the involvement of others, even statutory consultation, is often a condition for receiving grant aid for tree planting.

Access

Access routes both to and within forests are needed for road vehicles and those suited to less firm and rougher terrain for establishment, maintenance, and harvesting operations. Landing strips for aircraft are sometimes required.

Forest roads

The main function of forest roads is to provide access for timber harvesting vehicles. The construction of the final road network is sometimes delayed until shortly before harvesting begins though in such cases a reduced density of roads, often constructed to a lower specification, is made at the time of planting to allow access for labour and materials.

Detailed considerations in road planning by Johnston *et al.* (1967) are still relevant, though a more up to date account can be found in Hibberd (1991). The density required depends on the extraction equipment that is likely to be used and that, in turn, often depends on the terrain. Modern cable cranes, skidders, forwarders, and harvesters make it feasible to extract over much longer distances than when horses and agricultural tractors were used, but the standards of road construction and provision for good manoeuvrability are higher to cope with large, sophisticated vehicles. Decisions on desirable road networks cannot be made more than a few years ahead of felling because techniques are constantly changing: for example where a choice is possible, horse extraction favours roads at the bottom of slopes, whereas with cable cranes the top is preferable. Skidders have replaced cable cranes on moderately steep terrain and for them a road at the foot of the slope is ideal, for modern forwarders somewhere in the middle of a moderate slope may be better.

In deciding the spacing of roads, the aim is to minimize the total cost of the movement of timber from stump to market. This occurs when extraction costs per cubic metre from the stump to the road equal road construction and maintenance costs. It is therefore difficult to assess the best spacing and positioning of the final road network at the time of planting but an attempt is usually made to do this so that future road lines can be left unplanted. Roads are not easy to survey and may expose the forest to windthrow if made later by opening an established stand.

Additional factors that influence road planning are the provision of roadside working areas, linking with the public road system, and the interests of agriculture, recreation, and amenity. Where possible it may also be desirable to align roads through the least productive land and construct them in ways that minimize soil erosion.

The width left unplanted for roads must obviously be enough to accommodate the road itself and any associated drainage. It should also be wide enough to prevent excessive shading of the road surface which will deteriorate more rapidly if it remains wet. If windthrow, fire, sport, or conservation are additional considerations, wide road verges may be desirable.

Rides and tracks

Unsurfaced access routes are usually required in forests in addition to roads. They may be future road lines or simply routes left for tracked and other off-road vehicles, and for providing easy access on foot. They vary in width from rides 10 m wide to narrow tracks just wide enough for a vehicle.

The desirable density of unsurfaced routes has tended to decrease considerably as operations have become mechanized and also as the use of aircraft has increased. In the 1960s it was common over much of upland Britain and Ireland to leave one in every 12 to 14 rows unplanted to allow access for fertilizing by ground machines and eventual timber extraction (Dallas 1962). Since fertilizing is now normally done from the air and extraction can be carried out over stumps, these tracks are now obsolete, besides being unsightly and wasteful of ground (see Fig. 4.2, p. 53). Indeed in restocking no attempt is made to retain the old ride system and this often reduces the area taken up by rides and roads from 15 to about 8 per cent (Hibberd 1991).

A common practice is to surround compartments of about 25 ha with a ride, and to have some internal division as well, providing access to blocks of some 5 ha. Often there is about 1 km of access route, that may include surfaced roads, to about 15 ha of plantation.

Fire

The main features incorporated into plantation design to reduce the risk of fire spreading are the road system that provides access for fire fighting, and zones designed as relatively fuel-free barriers to the spread of fire (see Chapter 13).

Firebreaks are strips of at least 10 m wide that are kept clear of flammable vegetation by cultivation, herbicides, burning, mowing, or grazing. Today, they are made mainly on the outer boundaries of forests to prevent fires spreading in. Firebreaks used to be sited within forests as well but they can also act as channels, increasing the wind speed and causing severe wind turbulence. Indeed, they can exacerbate the severity of fires and increase, rather than reduce, the risk of fire moving across the forest. Breaking up large blocks of relatively vulnerable species with trees that do not themselves readily burn, such as many broadleaves, is now preferred. Trees along the margins of unavoidable internal breaks for access such as road lines are sometimes treated specially, for example by brashing, pruning, or heavy thinning, to minimize the risk of fire spreading across them. A fire may often burn freely through one species and age of crop but stops once it comes to another, less flammable, species or age class. Where fire risks are high, it may be desirable to avoid planting large homogeneous blocks.

However, these static measures are never the first line of defence—good relations with neighbours and the local community, rapid means of fire detection, and an equally rapid response capability, are far more important.

Stability

As noted on p. 187, where a forest is in an area of high windthrow risk careful design can reduce potential damage. This includes the special treatment of stand margins, particularly along internal road lines. Many of the features incorporated into protective firebelts will also contribute to stability.

Quality of water

The purity of water reaching streams, lakes, and reservoirs can be adversely influenced by new plantations and associated management operations. Firstly, on some sites with base-poor and weakly buffered soils surface water acidification can occur owing to the greater capacity of trees to trap air pollutants on their foliage. Although not the cause of the pollution, this can sometimes result in the tree crop mediating changes in stream water chemistry. Secondly, nutrient enrichment can occur from fertilizer addition that leads to eutrophication in lakes and rivers. In extreme cases eutrophication can lead to algal blooms that reduce oxygen levels in the water and can cause the death of fish and other aquatic life. Algae can block filtering equipment in water treatment plants. Thirdly, high levels of some nutrients, such as nitrates in drinking water, can be dangerous to health, though, compared with much farmland, forest cover gives rise to few nitrate-leaching problems. Fourthly, operations such as ploughing, draining, and road construction can increase sedimentation which can reduce water storage in reservoirs, block water courses, and damage aquatic wildlife.

These impacts can be largely eliminated by taking care over plantation design and paying due regard to operations that may cause damage. Road alignment and construction methods, and the frequency and slope of drains should aim to minimize erosion. A wide fringe of shrubs and suitable trees in a filtration or buffer strip next to watercourses should contain problems of acidification and nutrient enrichment (Forestry Commission 1993). Adverse effects are most likely at planting and immediately after clear-felling when soil disturbance is greatest and there is more rapid release of nutrients from decomposing vegetation and litter. The beginning and end of rotations are also times when sedimentation is most likely to occur. During the long period between canopy closure and clear-felling, forests are effective at retaining nutrients and conserving the soil and their acidifying effect is only of consequence where acid-critical loads might be exceeded on poorly buffered soils. If only small areas are planted or there is a wide distribution of age classes on a catchment, the risks are further minimized. In general, forest cover is far less polluting than agricultural land.

Conservation of wildlife

When, in the past, plantations replaced natural or semi-natural forests there was usually a decline in the diversity of plants and animals in the woodland. Where plantations are established on previously bare land, the reverse may be true. For example there has been an increase in the diversity of bird species in Britain where coniferous forests have been established on impoverished moorland, yet some species are also lost. Among birds, buzzards (*Bueto bueto*) use forests for refuge or to nest in but they need large areas of open land for hunting. Others, like skylarks, depend upon open habitats for both food and nest sites; they disappear as a tree cover develops (Moss 1979).

However, it needs stressing that today plantation establishment should neither be at the expense of natural forest nor on a site of intrinsic importance for conservation. International agreements concerning biodiversity (and the environment at large) have promoted greater awareness and imposed restrictions on how plantation forestry is used. Indeed, tree planting now contributes to creating new native woodlands with key biodiversity objectives (Rodwell and Patterson 1994; Peterken 1996).

In the sense that forests, managed primarily for the production of wood, are probably the greatest reservoir of wild plants and animals in a country, foresters have an important part to play in the conservation of wildlife. Quite small modifications to silvicultural practices can have surprisingly large effects (Steele 1972). There are few forests managed for timber production that cannot be improved as wildlife habitats. Ferris-Kaan (1995) and Peterken (1981) have discussed in detail the measures that might be taken to ensure biological diversity under free-living conditions. They include maintaining reasonable diversity of tree species, structural diversity in terms of age classes and layers in the forest canopy, maintaining special habitats, and protecting the forest from excessive disturbance.

Deer management

Though forests are the habitat of many animals, the main temperate species that have an impact on the design of plantations are deer. Since the 1980s, the many newly planted woodlands in lowland areas have provided ideal cover for deer and their numbers, in parallel with ungulates throughout the temperate world, have much increased. Deer must be managed to limit the damage they cause to trees and for the income that can often be obtained from hunters. They are also, of course, a form of wildlife enjoyed by the public and the conservation of acceptable numbers is an important aspect of management.

If uncontrolled, many species of deer cause considerable damage by browsing young trees, stripping bark, and fraying. They can prevent satisfactory establishment and cause diseases. In countries where numbers are excessive, they are regarded as extremely serious pests (König and Gossow 1979; Cooper and Mutch 1979). Attempts are made to keep populations to levels at which damage remains acceptable. It is quite certain that the presence of deer always entails additional costs. They should never, therefore, be introduced to areas where they are not already present. In Britain, red deer, roe deer, and, to a lesser extent, fallow deer (Fig. 4.1) and muntjac (*Cervus elaphus*, *Capreolus capreolus*, *Dama dama* and *Muntiacus reevesi,* respectively) are probably the second most important influences on the viability of forestry, wind being the first. In very big forests numbers can, to some extent, be anticipated by the proportion of forage to cover and its arrangement in space and time. These are the primary factors determining population levels (Thomas *et al.* 1976; Ratcliffe 1987; Ratcliffe and Mayle 1992).

Failure to take account of the ecological needs of deer populations and to design forest layouts that facilitate their management can lead to high and often ineffective

Fig. 4.1 Fallow deer, a species that is very damaging to European forests. (Crown Copyright.)

expenditure on fencing, and difficult shooting. Good forest design can minimize damage and enable any necessary shooting to be done relatively easily and even profitably.

Although knowledge has advanced in the fields of population dynamics, social behaviour, and interaction of deer species with their habitats, a detailed under-standing of what constitutes effective design of forests remains uncertain. Deer management is fraught with vested interests, folk lore, and emotion; legislation is frequently controversial, facts are short and difficult to obtain. For some species evidence from faecal pellet counts can predict deer numbers quite accurately and this can indicate the required cull, assuming the population biology is well under-stood. In Britain, cull levels recommended for roe deer have been virtually halved since the discovery that birth rates are much lower than were previously suspected (Ratcliffe and Mayle 1992)—this also indicates that the previously recommended cull levels of close to 40 per cent per annum had never been achieved!

In areas where roe and other species of deer are controlled effectively, provi-sions include leaving areas of at least 4 ha of favoured grazing unplanted and occa-sionally improving them by fertilizing. These areas may include indigenous woodland, the banks of streams, and roadsides (Prior 1983). Siting such areas on the edges of plantation blocks can be beneficial. Holloway (1967) stated that in Russia, elk damage is much worse in coniferous plantations mixed with *Populus tremula* than in pure conifers, while *Salix* spp. planted along water courses reduces damage to surrounding blocks of *Pinus*. Thinnings timed to coincide with periods of maximum food shortage can provide large quantities of alternative browse from the felled trees. Damage can also be reduced by leaving corridors of unplanted ground, or well brashed young stands, or old stands to enable deer to move between high and low ground. With some species of deer it can be important to incorporate traditional trails into the forest ride layout. Deliberately planting pre-ferred browse species of shrubs and trees, such as *Salix* spp., can help if they are sited carefully and the quantities are related to animal density. Also, among crop trees, some species are more able to recover from damage than others. *Pseudotsuga menziesii* tends to callous over bark-stripped wounds quickly and the wood does not become rotten, whereas *Picea abies* is more susceptible to decay.

Above all, culling must target the right animals. Certain individuals are more prone to strip bark than others. The elimination of these can reduce barking damage out of all proportion to the numbers shot. Dominant territorial bucks are often stabilizing influences on the population as a whole. They deter the immigra-tion of unattached males which can damage trees badly by fraying during displays in the mating season. Selective shooting can therefore result in a spectacular reduc-tion in damage whereas indiscriminate shooting can actually make it worse.

Fencing vulnerable areas is usually a last resort. It is costly and ineffective over long periods. Chemical repellents (Pepper 1978) and individual tree shelters of various kinds can be effective by preventing browsing on young trees, but are expensive and should only be used where small numbers of trees need protection.

In most plantation forests thought given at the design stage to the future management of deer can be a major step in limiting the severity of damage that is likely to occur. Unfortunately the trend to restructuring monotonous, uniform plantation to create more diverse and attractive forest also frequently favours deer. Their careful management is one of the many skills modern foresters responsible for multi-purpose forestry must now possess.

Landscape

Most people consider rural landscapes to be timeless, and resistance to any change is often very great. Thus, planting a forest on previously bare land, or subsequent felling of the same forest, can arouse public antipathy, and in areas of great natural beauty, can lead to outspoken criticism. An early example of this reaction is that of Symonds (1936) who criticized planting in the English Lake District.

> The planting is continuous, by the square mile. Where there was colour, this is first hidden, then dissolved; grasses, moss, plant-life perish as the trees form a canopy. Rock and scree, blue, grey or violet are still there, but hidden. What is seen is the rigid and monotonous ranks of spruce, dark green to blackish, goose-stepping on the fell side. Their colour is in effect one steady tone all round the dull year: there are no glories of spring and autumn for the conifer. Sunlight, that a broadleafed, deciduous tree reflects and vivifies, is annihilated on their absorbent texture: on a bright landscape they are so much blotting paper.

Recognition that ill-considered plantation establishment can mar the landscape (e.g. Fig. 4.2) led many forest authorities to consult qualified landscape planners from the early 1960s. Foremost among these in Britain was Crowe (1978) whose publications became valuable aids to an appreciation of landscape principles. Though there is a consensus of opinion about what constitutes an attractive landscape, it is less easy to analyse the components of a landscape and then design a forest that will conform to widely held views. Lucas (1983) discussed five major elements in landscape design; shape, visual force, scale, diversity, and unity (Fig. 4.3). The basis of successful integration is to understand the landscape and to design forests and operations within it in sympathy, so they appear as more natural features.

Landscapes are important in two senses. A forest must fit into the general pattern of the countryside, but the internal view, or immediate foreground, can also be important. The way a forest fits into the countryside increases in its impact as terrain becomes more hilly because the shape and composition of the woodland as a whole are more visible. The foreground view can be important in both flat and hilly country, especially in areas that are used for recreation. Here, diversity adds to the landscape: diversity in terms of species, tree sizes, open areas and dense woodland, and views out of the forest.

The Forestry Commission (1994) lists six key design principles, all of which can influence silvicultural and other decisions:

Fig. 4.2 Planting pattern commonly used in the United Kingdom in the 1960s. Here, a track is left between every 12 rows of trees to provide access for ground machines, especially for fertilizing. Helicopters have since made this unsightly practice entirely unnecessary.

1. Shape—appropriate shape is essential. Artificial geometric shapes and incongruities intrude into the landscape.

2. Visual force—forest shape should be designed to reinforce land form by rising up in hollows and falling on spurs and ridges to create a well-unified relationship between the two. This will often reflect the natural occurrence of woodland.

3. Scale—scale of a forest should reflect the scale of the landscape. Much depends on viewpoint. Blanket planting rolling over hill after hill can be seriously out of character with a locality.

4. Diversity—diversity of tree species, ages, stand composition, and other landscape features should be encouraged. Often careful species selection and matching species to changes in the site can do much to enhance the landscape. If done well, there is little conflict between good landscapes and good silviculture.

5. Unity—application of the above will help to create forest and woodland in keeping with the landscape and 'unite' it. Forests should usually blend into and not contrast with the landscape as a whole, and should possess complementary textures and colour.

6. The spirit of the place—well designed woodland will enhance and not detract from unique features of a locality. It will add to its specialness.

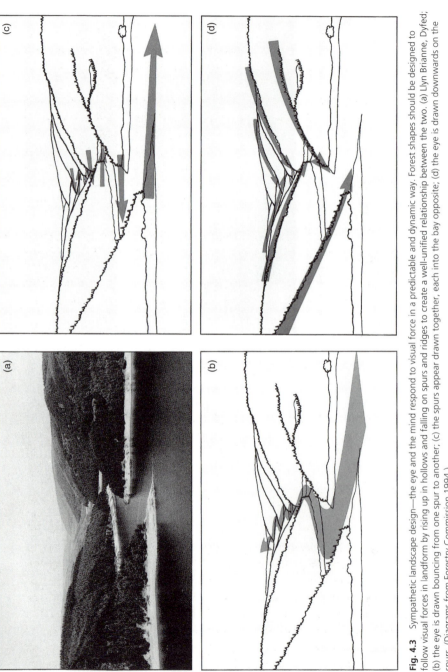

Fig. 4.3   Sympathetic landscape design—the eye and the mind respond to follow visual force in a predictable and dynamic way. Forest shapes should be designed to follow visual forces in landform by rising up in hollows and falling on spurs and ridges to create a well-unified relationship between the two. (a) Llyn Brianne, Dyfed; (b) the eye is drawn bouncing from one spur to another; (c) the spurs appear drawn together, each into the bay opposite; (d) the eye is drawn downwards on the spurs. (Diagrams from Forestry Commission 1994.)

Recreation

Of all outdoor activities, going into the countryside is the most popular and visiting forests is one of the attractions. There are numerous forms of recreation in forests, ranging from sightseeing, picnicking, walking, and camping, to educational and more energetic pursuits like orienteering and various forms of hunting. Each places different demands on a forest and on its design.

Usually the impact of recreation on timber production is insignificant since only very small areas are needed to accommodate large numbers of people and cars. In general, the features which make forests attractive for recreation are the same as those that make good landscape and a good conservation site: they include a large measure of diversity, mature trees, and the value of edges and water is high. One of the main criticisms of young conifer plantations is their close spacing. At very wide spacings, old conifers are seldom objected to but at close spacings the dark conditions in relatively young stands are usually considered both an eyesore and uninviting. They improve with time and thinning. The constraints on silviculture therefore usually include the need for heavier and earlier thinning than normal and retaining trees beyond economic rotation ages, at least immediately adjacent to popular picnic sites and viewpoints, and being more conscious of the need for fire protection. Access may need to be improved and people discouraged from going to certain areas. Ground-cover may need to be increased.

For planted forests where community use is uppermost, detailed planning, consultation, and integration with local interests is essential: see for example Agyeman (1996) 'Involving communities in forestry…through community participation'.

Forest normality and restructuring

In traditional forest management the term 'normality' means that a forest has an even distribution of age classes by area, ranging from newly planted crops to stands of rotation age. More precisely, normality implies arranging the forest so that production is the same each year, and can be maintained at a steady sustainable level. The concept can be extended to cover normality of annual production by species and size classes, of annual revenues, and of annual labour requirements.

Normality was considered very important in earlier centuries when local populations and industries depended on sustainable local supplies of wood because transport over long distances was impossible. However, some economists in recent times (for example Johnston *et al.* 1967) argue that normality has little relevance in places where cheap and efficient internal transport is available since no one industry or community depends upon a regular supply of material from a single forest. This is clearly true for large forestry enterprises that are able to even out supplies by felling in more distant forests. It is less true for private owners who depend upon woodlands for income, and the forests themselves are consigned to periods of considerable risk at one time from damage of various kinds, with possible economic consequences.

For many of the aspects of forest design discussed in this chapter and elsewhere, the measure of diversity that normality confers brings advantages, in addition to the conventional one of allowing a steady outturn of produce, accompanied by a steady income, and giving continued employment. For these reasons as the first rotation nears felling age, the opportunity can be taken to bring forward some fellings and delay others deliberately to restructure the plantations (Evans and Hibberd 1990). This often results in large plantation forests achieving greater 'normality' though not for the traditional silvicultural reasons.

Normality guarantees that the proportion of the forest at risk from serious damage by wind, fire, pests, and pathogens is at a minimum over the period of a rotation, since stands tend to be vulnerable to any one form of damage only at particular stages in their lives. Also, the creation of a varied forest cover ensures that conservation, amenity, and recreation interests are maintained as far as possible. Catastrophes occurring in forests that are not normal could have direct economic consequences, as well as causing less quantifiable damage to the environment. The main disadvantage of too strict an adherence to normality is inflexibility in management, but this is easy to avoid by, for example, considering age classes in 5-year cohorts, instead of annual ones.

It is often impracticable to spread planting of new areas over the whole or even a large part of a rotation but a satisfactory state, approaching normality, can be reached even without deliberate restructuring by the end of a second rotation, even in quite small forests. This is achieved by the varying optimum rotation lengths on sites of different levels of productivity, and of different species, the occasional need for some premature felling, windthrow, and the possibilities of some delayed felling.

Managing forests to maintain a mosaic of different age classes and, if possible, of species as well, should be an important aim of a silviculturist.

As Malcolm (1979) has pointed out, plantations cannot be expected to function as completely stable ecological systems in the face of environmental fluctuations, biological hazards, and changing demands that people make. Heterogeneity confers a measure of resilience that enables forests to persist and to continue to be productive, despite such fluctuations, and it also enables them to absorb changes in market demands, social attitudes, and levels of inputs more readily.

Proper design is vital if a successful multipurpose forest, of value and enjoyment to many, is to be created. A planted forest crop becomes less like a farm crop and more like wisely managed woodland where yields of timber are set alongside many other products and benefits.

Part II   Principles of plantation silviculture

5 Site preparation

One of the most important factors influencing the establishment and growth of a plantation is the soil. It must provide a suitable environment for the trees to anchor themselves, and to obtain the water and minerals needed for healthy growth. The provision of adequate site drainage and the correction of restricted soil profile drainage are two of the most significant ways of achieving long-term improvements in site potential for forestry. They usually benefit both growth and stability. The survival and growth of young trees are influenced by the conditions immediately around the newly planted tree. In addition to drainage, improvements may be achieved by surface cultivation, reducing competition for water and nutrients, as well as improving aeration and increasing soil temperatures for root growth.

The physical conditions for tree survival and growth can be improved on many sites in Europe. High precipitation in the west and north often results in water-logged, impermeable, and frequently peaty soils which are relatively unproductive before drainage. Seasonal flooding occurs along rivers and on flat land with high ground-water tables. Former agricultural soils are sometimes compacted and require cultivation. In fact, the successful establishment of trees is seldom possible without some kind of site preparation to ameliorate the constraints that would otherwise depress survival or growth or both (Sutton 1993). There are, of course, soils where site preparation is unnecessary or even undesirable. These include sedimentary soils that will dry too fast if the organic layer is removed.

Usually, soil manipulations that aim to achieve long-term site improvements are considered separately from those where the intervention has only a transitory effect. Long-term treatments include deep drainage, ploughing to disrupt ironpans, shallow cultivation on soils with thin peats to ensure roots grow into the mineral soil, and the restoration of abandoned opencast mining sites (see Chapter 14). Treatments that have short-term effects include patch scarification and disc trenching (Fig. 5.1) or mounding to improve conditions at the planting site during the establishment years.

Competing vegetation has a great influence on the water supply to seedlings. It can be controlled by scarification or the application of herbicides in the early stages of growth (see Chapter 8). Reduction of competing vegetation is often a major objective of site preparation. The competitive ability of natural vegetation differs from place to place: on fertile soils scarification and expensive weeding may be

Fig. 5.1 A scarifier working in Sweden.

needed. On infertile soils plantations often develop well without weeding, while on waterlogged sites it can be sensible to retain natural vegetation because it is needed as a water consumer.

Interactions between soil conditions and tree growth

Any form of site preparation has effects on:

(1) water movement;

(2) soil water content and its availability;

(3) soil aeration;

(4) soil temperature;

(5) soil texture;

(6) soil compaction;

(7) nutrient availability, especially nitrogen;

(8) competing natural vegetation and harmful insects;

(9) the cost of the whole plantation or (re)afforestation project.

Site preparation can also have a number of potentially serious detrimental effects, including increasing erosion rates, nutrient leaching, and inadvertent habitat destruction. It must, therefore, be meticulously planned and confined strictly to the parts of the site where it is necessary. Environmental guidelines for practice in Britain have been produced by the Forestry Commission (1993). They recommend aspects such as intensities and types of ground preparation, including the alignment and siting of drains and the treatment of buffer areas.

Regulating soil water content and its availability

Soil aeration

Soil must be aerated if roots are to grow satisfactorily. Gaseous exchange must take place between soil air and the atmosphere at a sufficient rate to prevent a deficiency of oxygen developing. Soil micro-organisms also respire, and under conditions of restricted aeration they compete for oxygen with the roots of higher plants. Excess water can result in anaerobic conditions developing and cause a lack of oxygen to tree roots while too little, or an irregular supply of water, results in reduced growth.

Intermittently flooded sites

The sudden flooding of a site can produce completely anaerobic conditions within a few hours as a result of gas displacement from the soil pores and the uptake by micro-organisms of the remaining dissolved oxygen. This can lead to a decrease or cessation of aerobic root respiration resulting in decreased growth, transpiration, translocation, and possibly the death of roots of some trees within a few days. In flood-sensitive species root tips are often damaged so that when aeration is restored, growth can only be renewed from regions nearer the stem. Even the most flood-tolerant trees must be flood-free for 55 to 60 per cent of the period when roots grow actively. In general, broadleaved trees show significantly greater tolerance to water-logging than conifers although the ranges overlap (Gill 1970). Death is hastened by toxic compounds that are produced by roots of some species in anaerobic conditions, including ethanol and ethylene (Coutts and Armstrong 1976). Anaerobic conditions also cause the soil itself to produce substances that are harmful to plant roots, possibly more harmful than the lack of oxygen. These include reduced iron and manganese compounds, and high concentrations of carbon dioxide and sulphides.

Tree adaptations to waterlogging

The mechanisms of tolerance of roots to waterlogging have been reviewed by Crawford (1982). They are commonly interpreted in terms of three mechanisms, two metabolic and one anatomical (Coutts and Philipson 1978*b*):

(1) tolerance to toxic concentrations of substances produced in the soil;

(2) metabolic adaptation of anaerobic respiration to produce non-toxic products;

(3) internal oxygen transport to maintain aerobic respiration and to oxidize toxic compounds in the rhizosphere.

The two most commonly planted trees on wet sites in Britain, *Picea sitchensis* and *Pinus contorta,* show marked differences in these respects. In a series of experiments, Coutts and Philipson (1978*a*, *b*) and Philipson and Coutts (1978, 1980) have shown that the pine will penetrate more deeply into soils liable to waterlogging. This is because, under constantly waterlogged conditions, large gas-filled cavities develop in the stele of pine roots but not in *Picea*. Similar enlargement of intercellular air spaces and relatively low soil oxygen demands are found in many other wetland species including *Populus* spp., *Salix* spp., mangroves, and *Pinus sylvestris* when grown under conditions of reduced aeration. These air spaces enable trees to transport their supplies of oxygen internally.

Oxygen enters the stem and roots from the leaves in some species. However, in most woody plants lenticels at the stem base or on roots in the top aerated soil horizons are more important entry points. On wet soils lenticels proliferate on parts of the root system above the water table. Other mechanisms may also operate. For example in *Pinus contorta* live roots have been found as far as 1 m below the water table, which is more than published distances for internal oxygen transport. Other anatomical modifications found in trees able to withstand flooding which may contribute to root aeration and ethanol removal include buttresses, pneumatophores, adventitious roots, and root branching.

Problems on fine-textured soils

In peats and deep, fine-textured soils such as clayey loams and surface water gleys, aeration is often reduced by an excess of water at the surface due to capillary action. Whether a tree is relatively flood-tolerant or not, root systems, even in drained areas, are commonly shallow and severely restricted at about 10 cm below the surface (Table 5.1) because of a lack of oxygen (Lees 1972). This results in

Table 5.1 Percentage of roots at different depths in the soil

Soil type	Root depth (cm)	Sitka spruce[1]	Scots pine[2]	Lodgepole pine[3]
		Surface water and peaty gleys	Deep peat	Deep peat
Country		N. Ireland	Finland	Scotland
		root dry weight (%)		
	Litter	58	70	62
	0–5	27		
	5–10	10	20	17
	10–20	5	10	12
	Below 20	0	0	9

[1] Adams *et al.* (1972); [2] Lähde (1969); [3] Boggie (1972).

poor anchorage and susceptibility to windthrow (see Chapter 12). Nutrients and, during dry periods, water may also be in short supply. On soils where rooting is not restricted by lack of oxygen or physical characteristics, Büsgen *et al.* (1929) quote rooting depths of *Picea abies* in excess of 2 m and of *Pinus sylvestris* to 5 to 7 m.

In regions with fine-textured soils that experience marked periods of drought, scarification can inhibit water infiltration and result in severe erosion due to soil capping.

Tree adaptations to drought

In many forest ecosystems water losses from leaves during the day exceed the amounts taken up by roots. Trees tend to become dehydrated as uptake lags behind transpiration. Temporary water deficits caused by excessive transpiration are not serious when the soil is well watered because leaves usually make up any deficit at night. Many trees, including temperate species, can also store considerable amounts of water in the sapwood that can be used during the day. However, as droughts intensify, water deficits develop that result in reductions in growth, injury, and ultimately death. The stress imposed by drought can also make unadapted species more prone to diseases and attacks by some pests. Shallow soils and deep, coarse-textured soils are especially prone to drought. South-facing slopes in the northern hemisphere can intensify problems.

Many species of trees exhibit various degrees of avoidance of drought damage through adaptations that retard water loss or increase water uptake. These include the ability to shed leaves to balance unavoidable losses of water. Leaf shedding occurs in dry summers. It was very noticeable among many deciduous species in western Europe during the summer drought of 1976. Other adaptations to reduce water loss include leaf waxes, which are common among the Mediterranean flora, and the presence of few and sunken stomata in many pines. The ability to close stomata quickly in drying conditions explains why pines will grow and thrive on dry sites while *Picea* spp., which do not have this ability, will not. Trees adapted to very dry climates have much reduced leaf areas and a temporary reduction is also shown by most species soon after planting, until their root systems are functioning properly. Where the problems of drought are predominantly those of a dry atmosphere, which causes high rates of transpiration, rigidly constructed evergreen leaves are best able to survive. Deciduous trees are adapted to periods of severe soil drought. Trees that grow well on dry sites usually have fast-growing, deep, and extensive root systems that develop at the expense of stem growth. This enables them to reach water that is unavailable to surface-rooting plants like grasses, and ensures the prevention, or at least postponement, of drought injury.

Soil temperature

The removal of ground vegetation to expose mineral soil increases the daytime soil temperatures during the growing season because more heat is absorbed by bare soil than by litter or vegetation. This warming effect increases if a mound or ridge is created up to a maximum height of 50 cm (Tabbush 1988). Soil on cultivated sites is therefore warmer in the growing season: soil temperatures cross the 4°C root growth threshold earlier in the spring, though they fall below it somewhat earlier in the autumn. In northern latitudes, and especially boreal regions, the benefits of higher spring time soil temperatures are important. To initiate the growth of new roots, a 5°C higher temperature is required than for normal root growth. Soil temperatures in the weeks immediately after spring planting are of fundamental importance for seedling establishment in the boreal zone (Söderström 1976).

Nutrient availability

When soils are poorly aerated and cold, the decomposition of organic matter and consequent release of nutrients are both slow. Nutrient availability can be markedly improved following mechanical site preparation for three reasons:

(1) the mulching and composting effects of inverted turfs or plough ridges;

(2) reduced competition for nutrients from weeds;

(3) higher soil temperatures and better aeration improve organic matter decomposition and hence the release of nutrient elements.

Effects of soil cultivation on damage caused by pine weevils and voles

Some insects and voles can be particularly damaging at the time of plantation establishment. The pine weevil (*Hylobius abietis*) causes severe economic losses by ring-barking and killing coniferous seedlings. Mechanical site preparation that exposes mineral soil reduces weevil damage (Christiansen and Bakke 1971; Eidmann 1979; Tabbush 1988) and is an alternative to chemical treatment where the weevil attacks are not too severe. In more serious cases, significant damage will be caused regardless of protective measures. Because the use of pesticides is being questioned in some parts of Europe, scarification has proved a useful way of reducing damage. By contrast, some leaf-eating insects like the larvae of the wasp *Acantholyda hieroglyphica* (Charles and Chevin 1977) thrive in the warm and sunny microsites in scarified spots or furrows.

Vole (*Clethrionomys* spp. and *Microtus* spp.) attacks are more seasonal but are often predictable and depend on the huge changes in population densities, especially in boreal zones. If the organic or grass layers are removed the ring barking and browsing damage they cause is less harmful.

Site drainage

The need for drainage to remove excess soil water is most common in peats, peaty gleys, and gleyed soils. The three major reasons for this are to improve tree growth, to increase the depth of rooting and so reduce the risk of windthrow, and to control runoff water from cultivated ground and thereby reduce soil erosion (Pyatt 1990).

The area covered with peat deeper than 0.3 m has been estimated at 450 Mha worldwide (Kivinen and Pakarinen 1981). Most peatlands in Europe are found in the north and west, and actual figures for each country are given by Paavilainen and Päivänen (1995). Finland has invested a great deal in converting peatland into fertile productive forests. Out of 10.4 Mha, 57 per cent has been drained to improve existing forests or establish new ones. Not all plantations on peat respond to drainage without fertilization as well (see Chapter 9). Research has demonstrated the extent of possible improvements in growth after drainage of different types of peats. Site classifications that predict post-drainage productivity, based on the natural vegetation before drainage, were published by Cajander as early as (1913), followed by Heikureinen (1979) and Heikureinen and Pakarinen (1982). For Sweden, Hånell (1988) has devised a useful system for the same purpose. It indicates that in the warmer parts of that country, growth responses to drainage can be as low as 0.5 m^3 ha^{-1} $year^{-1}$ on peats dominated by *Calluna vulgaris*, and as high as 11.4 m^3 ha^{-1} $year^{-1}$ on those dominated by tall herbs.

Principles of soil drainage

Any consideration of drainage must take into account various hydrological and related factors of a site. These include precipitation, infiltration capacity (which is the rate at which water will infiltrate from the surface into the soil) and soil permeability (which is the rate at which water will move through a soil profile), and evapotranspiration. The importance of these in forests has been discussed by Leyton (1972).

The infiltration capacity of a soil is determined by the characteristics of the soil surface. When mineral soil is unprotected by vegetation, the impact of raindrops may break down the surface structure to form a seal that, by retarding water entry, promotes surface flow and can cause flooding and erosion. A vegetative cover protects the surface from such deterioration and the litter and organic layers of forests can provide considerable additional protection. Generally, infiltration capacities in forests increase with the age and density of a stand. In mature forests they may reach maximum values of the order of 3 to 5 mm min^{-1} in wet soils and higher rates in drier ones.

However, infiltration cannot continue unless percolation through the soil provides space in the surface layers for the water, i.e. in soils where the number of large (non-capillary) pores is high, those with high organic matter contents, and where there is a good degree of aggregation in the surface soil. Characteristics that reduce the rate of percolation include high bulk densities and the presence of clay and silt in the subsoil. Most of these factors are interrelated and the superior properties of soils under forests are due to a generally higher non-capillary pore content as a result of the aggregation of soil particles by a high organic matter turnover. Forest cover therefore tends to improve the drainage characteristics, especially on fine textured clays and in the upper organically rich layers. Nevertheless, even after a relatively long period under forest, peats and clays in many places show poor drainage characteristics. They hold much water by capillary action and because of this, the movement of air into them may be extremely slow—often too slow to sustain tree roots. This is because water within capillaries is virtually immobile and can be removed only from a narrow band on either side of a drain (Stewart and Lance 1983). In some soils there may be no traces of oxygen at depths greater than about 20 cm, as was found by Armstrong *et al.* (1976) on a peaty surface water gley carrying a 23-year-old crop of *Picea sitchensis*. Drainage of these soils can, in fact, result in an increase in bulk density (Pyatt and Craven 1979).

Mull soils, which typically develop on base-rich sites, generally have better infiltration capacities than mor soils which are more characteristic on acid sites. Drainage of forest soils can easily be impaired by practices that destroy the surface organic layers or damage the surface structure, for example compaction by machines, burning, or grazing.

Largely because of increased interception, evapotranspiration from forests can remove 400 to 450 mm of water a year from a site in most upland areas of the British Isles. This is possibly twice as much water as is removed by open moorland vegetation, particularly in areas of high rainfall (Anon. 1976). Forests are therefore most effective at drying sites after canopies have closed. However, improved infiltration into the forest floor may lead to more sustained base flows from forested catchments.

Effects of drainage on tree survival and growth

Until the importance of drainage was recognized, most attempts at planting wet soils failed. There were many later failures too because the nutritional needs of crops·were not adequately understood.

A good experimental example of the effects of drainage on the growth of *Pinus contorta* is given by Boggie and Miller (1976). Their experiment, on a climatic peat, comprised five plots each of 3 × 30 m, isolated by perimeter ditches in which the water was maintained at different fixed levels. After fertilizing, trees were planted directly into the peat. The results 12 years later (Table 5.2) showed that where no drainage had been provided, mortality was high; the trees that had survived were, by chance, planted on hummocks. By contrast drained plots showed good or complete survival. All indices of growth including mean height and weight responded to the lowering of the water table, the lower the better.

These results illustrate the effects of lowering water tables alone but do not entirely represent forestry practice because, where drains are made, the material from them has to be put somewhere. It is usually spread in mounds or in continuous ridges alongside the drains. Young trees that are planted on top of these benefit not only from a slightly lowered level of soil water but also from a raised and well-

Table 5.2 Responses of lodgepole pine (*Pinus contorta*) on peat to drainage (from Boggie and Miller 1976)

Plot	Mean water table level (cm)	Survival (%)	Mean height age 12 (cm)	Weight of tree crop (t ha⁻¹)	Apparent yield class (m³ ha⁻¹ year⁻¹)
1	0.2	38	65	2.5	<4
2	10.9	97	125	12.8	4
3	18.8	100	129	13.0	4
4	24.4	100	219	52.3	8
5	33.6	100	305	96.8	10

aerated planting position. The more fertile condition of the 'sandwich' of decaying vegetation between the mound or ridge and ground surface also assists nutrition. There is considerable evidence (for example Binns 1962; Jack 1965; Savill *et al.* 1974) that in ridges the availability of nitrogen, and possibly other nutrients, is greater than in drained surface peat. This incidental effect of drainage has far more influence on tree growth than the lowering of water tables as such. In general the deeper the ploughing or cultivation, the more spoil is put on the surface and the better the growth of the trees. The additional growth reduces the time during which young trees are susceptible to hazards such as frost, grazing animals, and grass fires, besides having obvious economic benefits. With ploughing, roots proliferate along the ridges but are virtually absent from the furrows, and the uneven root-plate that develops can lead to serious instability later in life (see Chapter 12).

Drainage and cultivation practice

At drainage intensities that are practicable on climatic peats it is difficult to maintain water table levels much below 30 cm in summer or below 10 to 20 cm in winter, or to control them at all for a distance of more than about 2 m from the edge of a drain. No advantage is gained by deepening drains below 1 m on climatic peats because no measurable improvement is obtained in rooting depth (Taylor 1970; Burke 1967). However, on some lowland Finnish peats, Lähde (1969) found that for every additional 10 cm the water was lowered, the rooting depth was increased by 1 cm. On fine-textured surface water gleys, the problem is essentially similar. For example on gleys derived from carboniferous parent material in Northern Ireland, water table levels were scarcely affected by deep drainage except next to the edge of drains (Savill 1976).

For these reasons, in most wet soils, foresters no longer attempt to lower water tables over the site as a whole, as they attempted to before the middle 1970s. Instead, drainage schemes only aim to aerate the more permeable and better structured surface layers by increasing the rate of removal of surface water. This may be via plough furrows or other cultivation channels, and the elimination of particularly wet spots that might otherwise act as centres for windthrow. Most of the water carried by drains is therefore surface water or water from melting snow that can not enter the still saturated or frozen soil (Paavilainen and Päivänen 1995). It is usually collected from more steeply aligned plough furrows or other methods of cultivation (e.g. mole drains or ripping) into what are called cross or collecting drains.

The effect of drainage on the water regime of an ecosystem as a whole can be quite significant even if only surface water is removed. In a study of a catchment area in Finland, Seuna (1981) found an increase of 33 per cent in annual runoff during the first 10 years, when 40 per cent of the total catchment had been drained.

Planning drainage schemes

Site drainage involves laying out a permanent drainage scheme that will function for the whole of the life of a crop and even beyond. Drains either intercept groundwater, where the water table is near the surface, or collect surface water from soils with low infiltration capacities. To prevent erosion, the gradients of collecting drains should not normally exceed 3.5 per cent (2°) and must be even. Alignments are therefore just off the contour, at a slope that is steep enough to give an adequate rate of flow but not so steep as to promote erosion. The effective depth of collecting drains often needs to be 60 cm, implying an initial depth of 90 cm (Pyatt 1990) if some siltation and unevenness is allowed for.

Thompson (1979) gives a method of calculating the permissible maximum catchment area for collecting drains. It depends upon the velocity at which water can move without causing erosion after heavy rainfall. For most conditions permissible catchments vary from about 5 to 8 ha depending upon the type of drain constructed. In practice, they are often much smaller. Spacings between drains typically vary between about 25 m on very slightly sloping ground (< 3°) on peats, and 100 m on some mineral soils (Pyatt 1990).

Since about 1979, British practice has been to taper depths of drains and end them in buffer strips 5 to 20 m from water courses—the width of buffer strip depending upon the width of the stream. The intention is to allow the water to filter through the ground vegetation to prevent silting and consequent damage to fish spawning areas (Mills 1980; Forestry Commission 1993). This practice may also have considerable value on some shallow soils of low buffering capacity, reducing acid runoff into streams.

Before machinery was widely available, it was necessary to dig drains on wet sites by hand. This was done by cutting drains, usually between every fourth row of the proposed tree planting lines. The turfs cut from them were spread into lines at square spacings of about 1.5 m and a tree planted on top of each (4450 stems ha^{-1}). Because hand methods were so labour intensive, little afforestation was achieved until they were replaced by mechanical ones, such as modern mounding. The machinery for this was developed during and after the Second World War. By the early 1950s, machine drainage by ploughing or by excavators became the normal practice in places where it was needed.

Technical aspects of site preparation

The aims of site preparation vary according to site, climate, history, and the ambitions of the owner of the forest. Some treatments are carried out to improve survival and early growth of the seedlings; others are prerequisites for growth over the whole rotation.

Sometimes, on sites where large plants are used that have high root growth potentials at planting, no site preparation may be needed. This is particularly so if there are no special constraints to tree and root growth, such as wetness, compaction, or excessive weediness. Where these constraints exist, or where plants with low root growth potentials are used, some form of site preparation can be very beneficial. It is obviously important to ensure that the type used is appropriate to the site.

The variety of modern site preparation techniques is now so wide that methods are available that ensure successful establishment on almost any soil, as shown in Table 5.3.

Scarification

Soil preparation aimed at improving the microsite for seed or seedlings during early years after establishment by reducing competition from weeds and making nutrients and soil water more available, is termed scarification. It involves scraping off the vegetation, litter, and, most importantly, the organic layer to expose bare mineral soil or peat beneath. If, by contrast, the treatment involves incorporating surface organic matter into the mineral soil it is called tilling.

The most usual method is patch scarification, or exposing the mineral soil on a spot of 40×50 cm. Pine seedlings should, ideally, be planted in the centre of the

Table 5.3 Possible methods of site preparation

Mechanical site preparation		Other types of site preparation
Scarification	**Tilling**	
Patch sacrification	Power trenching	Mulching
Blading and scalping	Mounding with humus	Prescribed burning
Mounding without humus	Rototilling	Herbicides
Disc trenching	Ploughing	
	Bedding	
	Inverting	

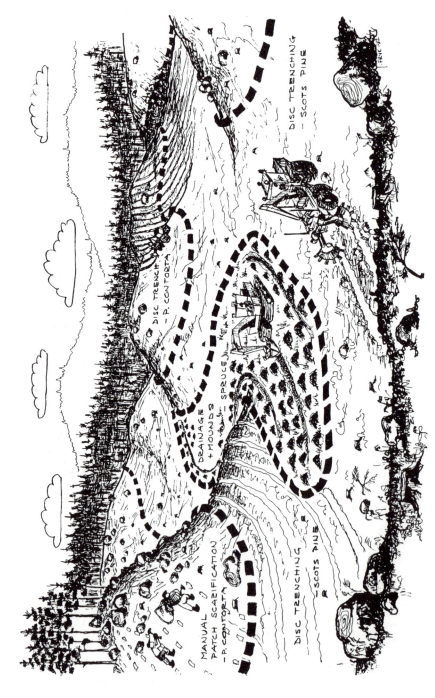

Fig. 5.2 An idealized example of site-adapted soil preparation. Though it does not make particular economic sense at present, it fits well with current ideas for improving biological diversity. (From a drawing by J. Fryk.)

spot but *Picea* does better if planted at the edge because its growth is stimulated by early root contact with the organic layer outside the cleared spot. On sites with high water tables a seedling planted in the hollow of the cleared area will have water-logged roots. In this situation an inverted turf beside the patch offers a more suitable planting spot. Where patches are extended into lines with associated furrows disced trenches are created. These offer a wider range of suitable planting spots.

Patch scarification can be extended into mounding if a volume of 0.1 to 0.2 m^3 of mineral soil is heaped on to the patch (see Fig 5.3). Planting is on top of the mound. Modern machinery can produce mounds either continuously or intermittently on disced trenched furrows. When tilling, the equipment can be set to mix the organic material with the mineral soil in varying proportions. Further details of these methods are described by Örlander *et al.* 1990.

On appropriate sites, trees established using these methods, and especially after mounding, compared to plantations established without any soil treatment have been shown to have improved survival rates and better growth in Sweden, south Finland, and Britain (Hämäläinen 1990; Tabbush 1988).

Scarification and tilling by hand used to be traditional for preparing planting spots for seedlings. This method is still the main one practised in Scandinavia in small clear cut areas and gaps, or when replacing dead plants. However, on a larger scale, the operations are now mechanized using a variety of heavy tractor-mounted equipment. Scarification by disc-trenchers can be carried out at a wide range of intensities to produce patches (Fig. 5.3) or furrows, sometimes with the same machine. On wet sites ploughs or excavators are used for simultaneous drainage and mounding.

Fig. 5.3 Site preparation methods for mineral soils, clockwise from top left—mounding with an excavator, disc trenching, patch scarification, motor-manual rotary cultivator, and manual patch scarification. (Drawn by Jarl Holmström.)

Where mounds are produced, they are usually about 25 cm high with gently sloping sides so that they are at least 50 cm in diameter at the base (Tabbush 1988). Mounds are particularly appropriate in areas of high windthrow hazard because the production of closely spaced furrows that restrict root spread is avoided (see Chapter 12). Mounds result in more symmetrical root systems and as a result stability is likely to be better. They are used to prepare sites for planting after harvesting many first rotation stands on peatlands in Finland, Sweden, and Britain.

Sites with a plentiful supply of capillary water to the surface tend to suffer seriously from frost-heaving of planted seedlings, especially in boreal regions with little snow. On these sites scarification should be avoided since it makes the problem worse. The retention of an organic litter layer can break the capillary water columns, or on waterlogged sites planting on a mound of mineral soil on top of the organic layer has the same effect. Mulching with sawdust may offer a cheaper alternative to this (Goulet 1995).

An extreme type of scarification is scalping as practised in the south-eastern United States. All debris from the harvest, including stumps and the organic layer, is removed with bulldozer blades, collected together in windrows, and usually burnt. The main reason for this drastic operation is to reduce future competition from weeds and to facilitate mechanized planting. A similar practice (Fig. 5.4) is also adopted in parts of Thetford forest in England where stumps and roots are removed to reduce damage from the butt rot disease *Heterobasidion annosum*.

Fig. 5.4 Complete clearance of a site prior to replanting at Thetford forest, England. All debris from the harvest, including stumps and most of the organic layer, is removed with bulldozers and collected into windrows. Stumps and roots are extracted to reduce damage from the butt rot disease *Heterobasidion annosum*.

A recent development in mechanical site preparation for harsh sites in boreal regions is described by Örlander *et al.* (in press). Their technique aims to create spots containing inverted surface organic matter covered by loose mineral soil, without making large mounds or ridges. Inverting appears to provide a suitable combination of reliable water and nutrient supply in a relatively warm rooting environment and avoids unnecessary seedling exposure to wind.

Ploughing

Two types of ploughs are in common use; single mouldboard ploughs, designed to produce single ridges and leave furrows that vary from 45 to 60 cm deep and 5 to 45 cm wide at the base, and double mouldboard ploughs which produce furrows of similar depth, but 35 to 100 cm wide (Thompson 1979, 1984). The material from these furrows is split into two ridges, one being placed either side of the furrow at a distance determined by the design of the plough. Trees are planted on top of the ridges which are relatively well aerated.

Modifications to ploughs can produce alterations to the shape of the ridge. Sometimes it is produced with a planting step to give early shelter and to enable roots to reach the more fertile ground of the vegetation 'sandwich' below the turf more quickly. Ploughs that produce deeper furrows and consequently bigger ridges are sometimes favoured because they result in better establishment.

Tunnel ploughing is a technique that has been developed in Ireland and has proved successful on many deep peats (O'Carroll *et al.* 1981). It produces a closed drain, the bottom of which is about 75 cm from the surface and recent models give drain channels 36 cm high and 26 cm wide. The material from the channel is extruded on to the surface like a conventional ploughed planting ridge but there is no associated open furrow. This method should result in more wind-firm crops with fewer nutritional problems since the roots exploit a much greater volume of peat. Tunnel ploughs are now widely used on deep peats in Ireland.

On mineral soils, such as gleys and peaty gleys, variants of the agricultural technique of mole drainage (Hinson *et al.* 1970) are employed to avoid the need for open furrows. They result in similar rates of tree growth to more conventional methods if competing vegetation is adequately controlled.

Cultivation of ironpan and indurated soils

Some soils originating from iron-rich rocks or young tills develop pans that are impermeable to tree roots and require cultivating to achieve satisfactory growth. Ironpan soils are common on heathlands in Europe. The pans at 30 to 40 cm depth can present impenetrable physical barriers to roots and to a lesser extent to water, possibly producing perched water tables. Equally impenetrable indurated zones may underlie the pans. Shattering them with ripping tines, or ploughing, can considerably improve aeration, drainage, and rooting. The mixing of the superficial organic horizon can also increase nitrogen and other nutrient availability. Bulk densities are reduced (Ross and Malcolm 1982) and oxygen levels in the soil are much improved by this form of cultivation (Pyatt and Craven 1979). Provided the pans are not too deep to reach with tines or ploughs, these soils offer excellent prospects for improvement.

The technical problems of cultivation were difficult to overcome until the development of machines capable of withstanding the rough usage to which they are subjected on these compacted and often stony soils. In Britain, the usual method used on ironpan soils is scarification of planting strips, combined with tining to break pans. On these sites, intensive or complete deep ploughing can result in greatly improved early growth due to the rapid mineralization of the mixed organic horizon. However, once the organic matter has been lost, serious nitrogen deficiencies may develop (Wilson and Pyatt 1984; Nelson and Quine 1990) unless fertilizer nitrogen is applied.

There are still a number of technical problems to overcome on indurated soils where impenetrable cemented horizons restrict water movement and root growth in a way similar to ironpans. The main problem is that the indurations are usually too deep for existing machinery, except at prohibitive cost. When indurations occur in fine-textured soils, rupturing them may give no drainage benefit. Rooting remains restricted because of the lack of oxygen in the soil.

Cultivation of drought-prone sites

By global standards, aridity is not a problem in most of Europe but periods of drought are common, especially in Mediterranean regions, and can be very damaging to young trees.

The most important measure that can be taken to prevent drought damage is to select species that are adapted to the anticipated conditions. In temperate climates, species of pines, and in warmer parts, eucalypts are almost invariably the choice for plantations. In many dry climates, tolerance to alkaline or saline conditions may be an additional requirement. A prerequisite for deep rooting is that compacted soils and soils with hardpans are loosened by deep cultivation. This is also very important for improving the storage of water in the soil. Shaping the terrain by terracing, to retain rather than to shed water, is a common means of cultivation in Mediterranean climates (e.g. Fig. 14.3, p. 208).

Site preparation with fire

Where debris creates a fire hazard, as in parts of British Columbia, Florida, and Mediterranean regions of Europe, prescribed burning is sometimes used for site preparation. It was also widely used in boreal parts of Europe for a period after the Second World War. Fire reduces the amount of competing vegetation, rodents, and some insect pests. Apart from the risk of it spreading to neighbouring stands, burning has the disadvantages of causing damage to the soil structure and increasing the leaching of nutrients. It may also increase the risk of some diseases, for example the fungus *Rhizina inflata,* and attacks by weevils, as well as making steep sites liable to erosion.

Plantations originating from prescribed burning are often fully stocked and uniform, indicating good survival. Comparative production studies sometimes indicate a slight reduction in growth during middle age. Pines normally do very well after burning but spruces grow better after other treatments. If well organized and if clear felled areas are large enough, the use of fire for soil preparation is cheaper than mechanical methods. After being ignored in silvicultural programmes for decades, deliberate burning is being used again as a means of site preparation for environmental reasons in some European countries. Many rare herbs, fungi, and insects are dependent on periodic fires. Controlled burning offers some protection against wild fires which is important in densely populated European countries.

Fire can be a threat as well as being useful and is further discussed in Chapter 13.

Costs of site preparation

Site preparation can be expensive and has to be economically justified through improved survival and growth over the whole rotation. Costs are partly determined by the size of the area to be treated. Generally, the smaller the area, the more expensive treatment becomes per unit area, because of the cost of moving machinery between sites.

The expenses can sometimes be offset by savings in other respects. For example when planting on ridges after ploughing, or on bare mineral soil after scarification, small and hence relatively cheaper seedlings can be used. Planting small containerized plants, where appropriate, is much faster than planting large bare-rooted seedlings. Weeding may not be necessary and improved survival can reduce or eliminate the need for replacing dead plants.

All operations in forestry are subjected to hazards from climate, pests and pathogens, unskilled labour, and political interventions in practises. If soil preparation is done well, the establishment period is shortened and risks of expensive failure are minimized.

Plant quality, plant handling, the timing and quality of planting, as well as pest control are all just as important as site preparation. Great financial penalties are often incurred by excessive short-term economies, and insufficient attention to the cost of the entire establishment operation (Tabbush 1988).

Environmental considerations in site preparation

When planning any site preparation, environmental considerations have to be kept in mind by proper analyses of the needs and likely responses of seedlings and trees.

The long-term effects of intensive site preparation are variable. Beneficial effects include the rapid establishment of the stand with deep root penetration and high litter production. Detrimental effects can include rapid mineralization leading to nutrient loss (Örlander *et al.* 1990).

Biological, technical, and economic advantages have to be weighed against environmental considerations. Drainage has an impact on water movement from forest land. This can affect the evenness of water flow in steams and rivers and can sometimes result in flooding far away from the forest. Leaching of nutrients after mechanical site preparation increases for a few years, causing losses and nutrient enrichment of rivers and the sea. The importance of these impacts has to be judged from experience and research in each region.

Any soil preparation has an impact on the original flora and fauna. Some species thrive on treated sites, including pioneers and species that depend on bare mineral soil. Others suffer, especially when the proportion of the area that has been treated is large. Scarification, for example, will kill a portion of the natural advanced growth. This may include seedlings with a potential to supplement the plantation with additional species and individuals. In other cases bare mineral soils may be necessary for establishing natural regeneration.

On a wider scale, soil preparation can have an impact on the carbon dioxide balance of the atmosphere. Draining, especially on peatlands can initiate a process of decomposition of these ancient stores of carbon, adding significant amounts of carbon dioxide to the atmosphere. This release is compensated to some extent by the growth of plantations that sequester carbon dioxide.

6 Choice of species

What species should I plant? This question leads to one of the most important of all decisions in plantation forestry because of the long cycles involved. An incorrect choice can result in poor health or growth, and even the loss of a crop. During the decades between planting and harvesting the requirements of the trees must be well matched to the chosen site. It is foolish to take risks with trees that may sometimes be justifiable with annual crops. The stresses imposed by droughts, gales, frosts, and fires, some of which will inevitably occur over a plantation rotation of 50 or more years, could lead to serious disease or pest outbreaks or even death. The species selected for planting should therefore be those whose requirements throughout life are likely to be satisfied by the site and climate in question. They must also fulfil the objectives of the planting scheme.

The process of species selection is done in three stages (Fig. 6.1):

(1) determining the characteristics of the planting site in terms of climate, soil, and other ecological factors;

(2) deciding which species and provenances are likely to thrive in such conditions;

(3) deciding which of one or more species, at the same time, satisfy the objectives of the planting scheme.

Site assessment

Climate

The climate of an area limits the range of species that can be grown successfully. Thus, obviously, tropical trees that require frost-free climates cannot be grown in cool temperate latitudes and though subarctic species might grow, preference is normally given to potentially more productive trees from regions with similar climates. Temperature, precipitation, and wind are usually considered the important elements of climate for forestry.

The more favourable the climate, the greater is the range of species from which a choice can be made. In upland regions where the climate is harsh and soils are often poor, the number is limited to two or three whereas, on good soils, in the better climate of the lowlands, a choice from about 20 may often be possible.

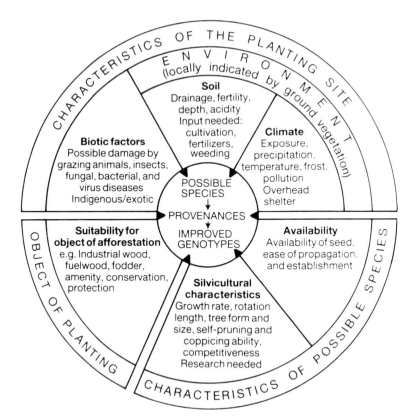

Fig. 6.1 Considerations in selecting species.

Characteristics of the climate of Europe

Europe has a remarkably favourable climate in comparison with those in other, similar, latitudes. Its winters are very mild and summers are warm, but not excessively hot. The range of temperature is small for the latitude and rainfall is abundant and well distributed. These conditions arise because of the extensive and widely dispersed marine influences that ameliorate the climate. Also, the absence of any significant mountain ranges along the west coast (as there are in both North and South America) is a feature that confers great climatic advantages. Unlike North America, the coastal regions of Europe do not receive a superfluity of rain and have arid lands to the leeward side which experience extremes of temperature. The main European mountain ranges (Alps, Pyrenees, Carpathians, and Caucasus) have east–west orientations that result in much more gradual rainfall and temperature gradients from west to east. They also block excessive latitudinal movements of both tropical and polar air masses. As a result, the climatic transition is gradual from the oceanic seaboard to the continental east, where winters are colder and the range of temperature greater (Kendrew 1961). Most boundaries are arbitrary, except for the

Mediterranean region which has produced a characteristic type of vegetation, typified by olive trees. Its distinctive climate shows much more periodicity than the climates in the rest of Europe, with mild wet winters and hot, dry, cloudless summers. Fires are more serious in the Mediterranean region than elsewhere (see Chapter 13).

In Europe, two broad patterns of variation exist at right angles to each other. One is latitudinal, from south to north, with decreasing temperature as the main influence on growth. The other is an increase in oceanity from east to west. The two gradients combine to cause a general increase in severity of conditions for plant growth in a north-westerly direction, reflected in the downward shift of altitudinal zones of vegetation in the same direction. Tree lines in Britain and Ireland are often lower than 400 m, but they rise progressively across Europe to over 2000 m in the Alps. The ecological and physiological reasons for the occurrence of timber lines have been discussed in detail by Tranquillini (1979). At high elevations, most heat received by the soil and vegetation is by direct insolation rather than indirect warm air currents. Hence, there is a marked contrast between the vegetation of north- and south-facing aspects at the same elevation.

Broad climatic patterns influence many regional approaches to species selection and other aspects of forestry. Climatic variations on a much more local, even site, scale can also have a considerable bearing on the selection of species.

Frost

Frosts during the growing season can cause extensive injury or death to young trees of many species. In older crops frost can kill foliage, slowing the rate of growth and result in forking and crooked stems, though death of the whole tree is unusual except in very sensitive species.

Sites where frosts are most serious are hollows, especially on otherwise relatively level ground, where denser, colder air collects. On such terrain, the depth of freezing air is often under 1 m and seldom exceeds 3 m, hence the most serious damage is usually confined to young trees (Day and Peace 1946). Many species become hardy to low temperatures during autumn, and midwinter temperatures as low as −40°C may do no damage. Hardiness is achieved through physiological mechanisms during the dormant period. They prevent water freezing within certain crucial tissues, either by supercooling or by moving water from cells so that extracellular ice forms (Levitt 1972; Burke *et al.* 1976). The same species may become susceptible to severe injury by frosts of only −2°C in spring and many are specially at risk during the early period of bud burst and extension. In most of Europe, susceptibility to damage rises until the end of May or early June. From then it remains reasonably constant until September or October, though differences of timing and degree occur between species and races.

A marked feature of many northern temperate climates is their variability over short periods of time. In upland regions of Britain, temperatures commonly oscillate across the 6°C growth threshold, especially in spring and autumn. This causes intermittent growth. They frequently cross the 0°C ground frost threshold in winter and at night occasionally throughout the rest of the year (Taylor 1976). Evergreen trees such as *Nothofagus procera* and *N. obliqua*, *Pinus radiata*, *Cupressus macrocarpa*, and some *Eucalyptus* spp., that are adapted to mild winters, can sometimes grow continuously in mild places such as Brittany, the south-west of England and south-west Ireland but they may be killed in rare, severe winters.

Precipitation and evaporation

In western Europe, precipitation is generally abundant, and in the mountains excessive in all seasons except spring. The driest areas are the Paris and Garonne basins, The Netherlands, Belgium, north Germany, and the east of Britain, with from 500 to 750 mm of rain a year. Even low hills have significantly more than neighbouring plains. The wettest areas, often with well over 2500 mm, are the western seaboards of France, Britain, and Ireland and some other upland parts of Britain.

In the low rainfall areas droughts may occur in summer. Though the frequency of rainfall is normally sufficient to prevent drought damage, especially in deep-rooted trees, some species are particularly sensitive to water stress. They benefit greatly, when young, from good weed control that reduces competition for water. Much of the reduction in growth potential can be accounted for by the effects of high vapour pressure deficits causing stomata to close and halting photosynthesis.

Soil

In Europe, forestry has traditionally taken second place to agriculture, and most forest sites are marginal for agricultural production. Apart from physical problems like steepness, stoniness, and exposure, many forest soils tend to be drought-prone or waterlogged, or suffer from extremes of acidity or alkalinity, or have compacted or cemented layers. In Britain, for example, two-thirds of the state-owned forests have soils which suffer from poor aeration due to permanent or periodic waterlogging (Toleman and Pyatt 1974). There are many important interrelationships between the physical, chemical, and biological influences in a soil that can influence species choice.

Soil depth

A good potential rooting depth is one of the most important determinants of productivity in trees. Shallow soils and those with iron pans or cemented layers that restrict rooting can lead to instability, poor nutritional conditions, serious problems of lack of water in dry periods, and waterlogging when it is wet. The exception is on shallow soils over deeply fissured bedrock where growth can sometimes be satisfactory. Particular difficulties arise in limestone country and in hot dry regions where calcareous pans have developed and soil erosion has subsequently taken place. These conditions are widespread in the Mediterranean region.

Soil structure and texture

The structure of a soil is defined by the spatial arrangement of the soil particles. Water retention and movement, aeration, bulk density, and porosity are all influenced by the overall aggregation or arrangement of the particles and are discussed by Pritchett (1979), among others. Structure can be altered to some extent by drainage and cultivation to improve rooting depth. Soil texture, by contrast is not easily changed. It is described by the proportional distribution of sand, silt, and clay particles. The presence of organic matter and enough biological activity in the soil to aid aggregation into a satisfactory structure is important. These factors have major bearings on how different species will grow.

Slowly permeable soils can prevent the infiltration of water and air to any significant depth. Oxygen levels, except at the surface, are often too low for tree roots to survive. At the other extreme, excessive permeability can lead to rapid movement of water through the profile and drought conditions may develop. *Picea* spp., for example, are well adapted to surface rooting on wet, heavy, relatively structureless soils while pines will tolerate drought. A few species, such as many *Populus* spp., are very exacting, requiring high fertility and ample, well oxygenated soil moisture at between 50 and 100 cm from the surface and not further from it than 150 cm (Jobling 1990). Many other species thrive on a wide range of soils.

Soil fertility

Although adequate soil fertility is critical, forests recycle nutrients efficiently, as described in Chapter 9. Most species will grow best on fertile soils, but it is nevertheless relatively easy to correct deficiencies by fertilizing. The capacity of soils to retain nutrient ions, particularly cations, in a form available for plant use is of particular importance on sands and other soils with very low nutrient reserves.

Many broadleaved trees differ from conifers in that they are relatively unable to obtain nutrients from intractable soil minerals. Most broadleaves have been described as site demanding by Miller (1984), and are more commonly associated with soils where nutrients are present in relatively easily available forms. A few species, such as tree lupin (*Lupinus arboreus*), are extremely valuable pioneers (Marrs *et al.* 1982) being able to extract phosphorus from sources unavailable to most other plants, and then make it available through the recycling of litter.

Soil reaction

Though soil pH varies slightly with season, and often considerably with depth, most soils on which plantations are established vary from highly acid, with a pH of 4.0 or less, to very basic, with a pH of 8.0 or more. Some species are demanding in their tolerance to soil pH and are quite incapable of growing well on soils with the wrong pH. In general, forest trees are well adapted to acid soils. Most conifers grow best where pH ranges from 4.5 to 6.5, though many broadleaves appear to have an optimum range near neutrality. The few trees that thrive in more alkaline conditions, such as Austrian pine (*Pinus nigra* subsp. *nigricans*), *Fraxinus excelsior,* and *Acer pseudoplatanus* have special uses on some sites.

Other ecological factors

Species selection can be influenced to varying degrees by many other factors unrelated to soil or climate, including:

1. The predisposition of a species to damage by large grazing mammals—sheep, for example, seem attracted to various silver firs (*Abies* spp.) which, like many shade-bearing trees, are slow at making early height growth and can suffer badly. Deer, squirrels, rabbits, and other animals can also cause damage that may influence species choice if no special protection, like fencing, is provided. For example the prevalence of grey squirrel damage on *Fagus sylvatica*, *Quercus*, and *Acer pseudoplatanus* in England is one of the major constraints to growing these species.

2. Atmospheric pollution of various kinds can also damage some species more than others—see Innes (1993) and Schlaepfer (1993). Usually, evergreen conifers with high leaf area indexes are most susceptible to pollution.

3. Susceptibility to damage by fires can limit the choice of possible species on certain sites.

4. The care the trees will get immediately after planting is important. If weed control is unlikely to be thorough, it may be better to plant a competitive tree like *Larix*.

5. It is also possible to assess many of the more obvious risks from insects, fungi, viruses, and bacteria that have been discussed in Chapter 2.

Potential plantation species and limitations in availability

Outside the Mediterranean parts of western Europe, there are about 40 native species of trees. In the same region, Mitchell (1988) states that over 500 species can easily be encountered by anyone looking in parks and gardens. If special collections with associated varieties are also included, the number, in Britain at least, rises to about 1700. One may wonder why, out of this huge variety, no more than perhaps a dozen are commonly encountered in European plantations. The answer is that many are not sufficiently hardy, or have such special site requirements, or grow too slowly to be considered for the sites generally available. Others are unsuitable because sufficient viable seed is not available or they are difficult to raise in nurseries. Some species are unsuited to growing in plantations, and may suffer from pests and diseases if grown in what is almost inevitably close to a monoculture. The great majority are ruled out because they do not provide the sort of product required.

The Macedonian pine, *Pinus peuce*, for example, is a potentially useful species were it not for problems in the nursery and establishment phases. Seed is difficult to obtain from its native Balkans, and Lines (1985) states that it has a tendency to poor or delayed germination, which is partly due to incomplete embryo development. After planting, early growth is very slow and bushy until the fifth or sixth year. Another potentially valuable pine for some soils is the north American *Pinus strobus*. It is not planted because it commonly suffers from fatal infections of the blister rust *Cronartium ribicola*. Beech (*Fagus sylvatica*) is unsuited to plantations in parts of its range, where growing season frosts on open ground can cause high levels of mortality, unless shelter is provided. Others like ash (*Fraxinus excelsior*) and cherry (*Prunus avium*) do not thrive in extensive monocultures because disease and pest outbreaks become damaging. *Betula pendula* could be grown much more widely than it is if adequate markets existed for its timber, but at present these exist only in Scandinavia.

An increasingly emotive issue is whether the species planted should be indigenous or exotic.

Indigenous species

An indigenous species is one that is native to the country concerned, though not necessarily to all parts of it, and no species is suited to all sites (Evans 1992). In regions of the world containing good indigenous tree floras, including a range of species that meet industrial and other demands, it is rare to rely upon exotics since few prove widely useful. This is true, for example, of the south-eastern United States and the Pacific north-west (Zobel *et al.* 1987). Indigenous species have the advantage of being well adapted to the local climate and potential hazards, such as pest outbreaks, through long periods of evolution and coevolution. For ecological reasons, current thinking everywhere is that native species should be favoured over exotics if possible.

In comparison with many other temperate regions, Europe has a small native tree flora. The generally east–west orientation of the main mountain ranges that give the continent such a favourable climate is also partly responsible for its relative paucity of species. During recent ice ages they acted as barriers to the retreat of the flora to the south and prevented invasion as conditions improved. Consequently many species have been lost. This is believed to be the explanation of the richness of species in North America, where mountains have a north–south orientation, compared with northern Europe (Zobel *et al.* 1987). Europe has only five native conifers north of the Alps compared to at least five times this number in western North America. More regionally, Britain has only one commercially valuable native conifer, *Pinus sylvestris*, and Denmark and Ireland have none.

If an indigenous species will grow well on an area to be planted and will also provide the sort of produce required, there is seldom a compelling reason to widen the choice. This is well illustrated, even in the relatively impoverished continental European conditions, by the widespread planting of the indigenous *Picea* and *Pinus*, as well as broadleaves like *Quercus*. Exotics are comparatively rarely used except where the native tree flora is deficient, as in Ireland, Britain, and Denmark, or where an exotic fills a niche for which there is no suitable native tree, as with *Pseudotsuga menziesii* in much of continental Europe, which produces such high quality softwood.

Exotic species

An exotic is a species that is grown in an area where it does not occur naturally. The benefits and risks attached to growing exotics are discussed in Chapter 2. In regions where the native species are relatively unproductive or otherwise unsuitable, foresters usually rely upon exotics because of their greater levels of productivity (see p. 30). For northern Europe, by far the most important donor region is western North America from which *Pseudotsuga menziesii* (Fig. 6.2), *Picea sitchensis*, *Pinus contorta,* and several other, less important, conifers originate. Conifers are normally favoured as exotics because of the wide demand for their timber, their high productivity, adaptability, hardiness, and the ease with which they can be established. Except for eucalypts, which are widely planted as exotics in warmer temperate regions such as Portugal, and to a lesser extent *Populus* spp., exotic broadleaved species are much less successful.

The practice of introducing trees into new regions of the world is as old as civilization itself. However, it is only since the middle 1800s that the formation of plantations, particularly of exotics, has been regarded as specially important. The period since the First World War has seen a rapid increase in both area and number of species planted.

The history of many introductions has been described by Streets (1962) and Zobel *et al.* (1987). They frequently follow a pattern of initial establishment in gardens and arboreta and then, if successful, perhaps half a rotation later, these are followed by small-scale planting on estates and by government departments. Large plantations of the most promising species are established some 20 to 50 years after that. *Picea sitchensis*, which is the major plantation tree in Britain and Ireland, was introduced in 1831. It was established as one of the chief exotics in the early 1920s and by the mid 1950s it became the most widely planted tree. The history of *Pinus radiata* in New Zealand follows a similar pattern, being introduced about the end of the 1850s and planting started on a vast scale in the 1920s. Thus, a long period of trial, amounting to at least one and a half to two rotations is usually thought advisable before an exotic is widely planted. Introductions are still taking place and there is still no substitute for extended trials before major planting schemes are started. If there is no time for such trials, the decision makers must accept the risk that their investment might not yield a financial return.

Genetic resources

The term 'genetic resources' refers to the gene pool of both wild and cultivated populations of a species. Populations of trees have high levels of genetic diversity

Fig. 6.2 A 30-year-old stand of Douglas fir (*Pseudotsuga menziesii*), a commonly used exotic from North America, growing in France. (Photo: by INRA.)

relative to other plants, a consequence of both their biological characteristics and generally short history of domestication. The genetic resources of currently or potentially important plantation species are widely distributed: many are represented primarily in natural populations, and are threatened by forest loss or degradation (e.g. National Research Council 1991; Sayer and Whitmore 1991). Existing plantation forests also represent a genetic resource, although sometimes one of limited value because of a narrow genetic base (e.g. *Eucalyptus globulus* in Portugal—Pereira and Santos Pereira 1988; *Pinus radiata* in Australasia—Moran and Bell 1987). The current availability of genetic resources varies greatly, therefore, with species.

The genetic resources appropriate for particular plantation forestry activities will depend on:

1. The purposes for which afforestation takes place—there is enormous genetic variation in most characteristics between and within most species. In the case of simple plantation forestry, wood production and quality traits are likely to be of greatest interest; in the case of more complex plantation forestry, a range of traits such as the interaction of the species with crops, the palatability and nutritional value of its foliage, fruiting characteristics, the calorific value of its

wood, or its ability to coppice from stump, may be of greater significance (e.g. Boland 1989).

2. The environmental characteristics of the sites available for establishment—typically, sites available for plantation establishment have been those either not favoured for agriculture or degraded by overexploitation; physical and nutritional soil characteristics have therefore been poor; sites may also be more subject to climatic extremes. Consequently, the range of species which can be successfully established may be limited, at least initially. In such cases, simple plantation forestry may be a necessary precursor to the development of more complex ecosystems (e.g. Gilmour *et al*. 1990).

Provenance

When dealing with both indigenous and exotic trees, it is not sufficient to decide simply which species to plant without considering the original geographic source of the seed as well—that is, the provenance. Some of the worst mistakes in plantation forests are the result of well-intentioned foresters importing seed from stands that look good but which are quite unsuited to the new environment (Lines 1967). For example, in the north temperate zone, seasonal fluctuations in day length and temperature control growth periodicity. The northward movement of *Betula* spp., which are strongly adapted to their native photoperiod and temperature conditions, results in the risk of damage by frost because late-season growth does not stop soon enough. Movement of trees southwards or from continental to maritime areas results in early flushing and spring frost damage (Habjørg 1971). Similar problems are found with many other species, though in some cases movement can be beneficial. For example in northern Sweden, *Pinus sylvestris* originating from 2° latitude further north than the planting site is normally used in order to withstand the harsh environments on clear cut sites. The trees grow slightly more slowly than the indigenous provenance, but survival is better.

The nature of variability within a species has been discussed by Callaham (1964), Burley (1965), and Perry (1979). There are always genetic differences between individuals of species that reproduce sexually, and differences also occur between local populations, which are often referred to as provenances. Population differences occur because the individual members of a population tend to breed with each other more than with members of distant populations. Genetic differences may be expressed in different morphological or physiological traits. They arise because, under the pressures of natural selection, genotypes that are unsuited to the local environment do not survive, whereas the progenies of more fit individuals do. The population, therefore, becomes increasingly suited to the local environment for such characters as adaptation to photoperiod and drought, frost resistance, and growth, besides less obvious features including mycorrhizal associations and enzyme production.

The extent of genetic differentiation between populations depends on the natural range of the species and the variability of the environment through the range. In most widely distributed species, patterns of genetic variation reflect patterns of environmental change, and discontinuities in one are commonly associated with changes in the other. Thus, in species where there are no distinct populations but a continuous intergradation, there is a gradual change in the character under consideration. Ecotypic variation is the alternative and may occur between geographically separated populations or populations on markedly different soil types.

It is widely accepted that the pattern of variation and growth within a tree species is continuous and is closely related to climate (Burley 1965). Individuals of a widespread species therefore show a continuous variability in response to photoperiod, temperature, and precipitation. Superimposed on this continuous pattern, discontinuities, such as abrupt changes in soil type, may lead to recognizable ecotypes. The selection of a suitably adapted provenance can therefore be critical when planting an exotic species with a widespread natural distribution, and a knowledge of the patterns of natural variation is of considerable practical importance.

Some species are relatively plastic in that they have a good ability to adapt to conditions that differ from those of their native habitat. *Pinus radiata* and *Robinia pseudoacacia* are among the best known plastic species. *Pinus radiata* has a native distribution totalling only about 7000 ha in California and on two Mexican offshore islands, yet it is more widely used as an exotic than any other species. Species which lack plasticity but which have wide natural ranges, such as *Pseudotsuga menziesii* and *Pinus contorta*, also prove successful, but only if the right seed origin is selected for each region.

The forester has much to gain by using the best possible seed source for raising trees for plantations. It is a false economy to buy seed of an inappropriate provenance, however cheap and readily available. By contrast, seed sources of native species that have been undisturbed are usually the best adapted for survival and reproduction in the local conditions. They are not necessarily the most productive, however. One of the best known examples of this is *Picea abies* in Norway and Sweden. After the last ice age this species migrated to southern Norway around the Gulf of Bothnia (Lagercrantz and Ryman 1990), in a southerly direction, and on the way became adapted to the relatively harsh conditions further north. Nonnative, more southerly provenances that would have reached there had the Baltic sea not prevented their spread often grow faster on some sites (Giertych 1976).

Provenance differences are best determined by experiments carried out in the region where the tree is to be planted. If the species exhibits continuous variation, it is theoretically possible, though seldom done, to predict the performance of an unknown provenance from that of two provenances at opposite ends of the cline. In ecotypic variation, prediction is restricted to within the limits of the ecotype.

Picea sitchensis is a species that exhibits continuous variation and is widely planted as exotic in Britain, Ireland, and parts of France and Scandinavian countries. It originates from an 80-km-wide zone at low elevations along the Pacific coast of the United States and Canada, spanning more than 22° of latitude, from Alaska to California. In Britain, experiments indicate clines of decreasing vigour of growth, and increasing resistance to frost damage among provenances from Oregon to Alaska, while Californian provenances generally do not survive. The choice of seed origin must attempt to compromise between these two characteristics. Seed from the Queen Charlotte Islands in British Columbia satisfactorily combines hardiness with adequate growth rates and often gives the best height growth on all sites (Lines 1987).

Lodgepole pine (*Pinus contorta*) is another exotic planted quite widely in parts of Europe. It exhibits both continuous and ecotypic variation. It has a much greater natural distribution than *Picea sitchensis* in north-west America, spanning some 33° of latitude, from Alaska to Baja California in Mexico, and 33° of longitude, from the Pacific coast to South Dakota. It covers a very wide variety of climatic and soil conditions. At least three interfertile races of *Pinus contorta* are usually recognized as subspecies that differ markedly in ecology and morphology. Subspecies *contorta* grows along most of the Pacific coast in bogs, on sand dunes, and on the margins of pools and lakes. It is short, shrubby, and of poor form. This gives way to subsp. *latifolia* in the intermountain systems from central Yukon to eastern Oregon and south Colorado, and to subsp. *murrayana*, mainly on the Cascade and Sierra Nevada mountains in Oregon and California (Critchfield 1957). The last two have better formed, slender, tall straight trees. The coastal subsp. *contorta* gives much the highest yields as an exotic in Britain and Ireland and is very tolerant of exposure. A constant dilemma has been how far to compromise on form (Fig. 6.3), while retaining vigour and hardiness when the species is grown for timber. In Ireland the most vigorous provenance of subsp. *contorta*, from Oregon and Washington, is almost exclusively planted (O'Driscoll 1980). By contrast, in Britain, slower growing, more northerly coastal provenances of subsp. *contorta* are recommended (Lines 1996) besides some inland provenances of subsp. *latifolia* and intermediate ones, such as from the Skeena River. By contrast, in Sweden, inland provenances of subsp. *latifolia* from British Columbia and Yukon are used.

Zobel *et al.* (1987) have provided some general rules about matching provenances to sites as exotics:

1. Do not obtain seed from high-elevation or high-latitude sources to plant in low elevations or low latitudes.

2. Provenances from a maritime climate should never be used in a continental climate.

3. Provenances from milder climates usually grow well unless there is a risk of freezing.

4. Do not move trees from areas of uniform climates, where minor fluctuations in rainfall and temperature occur, to those with severe and large fluctuations.

5. Do not choose trees that originate in basic soils for growing on acid soils or vice versa.

(a) (b)

Fig. 6.3 *Pinus contorta* from (a) Long Beach, Washington (subsp. *contorta*) and (b) Smithers, Skeena River (subsp. *latifolia*) indicating the differences in form of these two seed origins.

Genetic improvement

There is also considerable variation within any natural population that cannot be attributed to differential selection along environmental gradients: this may often be greater than responses to environment. Experiments frequently show that within-provenance variations are larger than those that occur between provenances. A recognition of this type of between-tree variation is the basis of most tree breeding for increased productivity, improved stem form, disease resistance, and other attributes. Over the last 20 years, access to genetic markers has increased enormously and has revolutionized approaches to measuring genetic variation. Müller-Starck and Ziehe (1991) and Kremer *et al.* (1993) have reviewed current knowledge of patterns of genetic variation in many European forest tree populations.

Genetic improvement of pure or hybrid species is a feature common to all successful plantation programmes and approaches to them have been discussed by Zobel and Talbert (1984), Kleinschmit (1986), and Tessier du Cros (1994) among others. It is helpful to conceptualize forest genetic and tree improvement activities in terms of the three phases identified by Cheliak and Rogers (1990):

1. Conservation—in addition to efforts at genetic resource conservation *per se*, it is also possible and desirable to design breeding strategies which specify the maintenance of genetic diversity as an explicit objective.

2. Selection, breeding, and testing, based on the recurrent cycle described by White (1987) and Cotterill (1986) who demonstrate that, in most cases, relatively simple methods can be as effective as more complex alternatives.

3. Propagation, that is the transfer of gains realized by genetic improvement from the breeding population to operational production—currently feasible propagation options are the production of genetically-improved seed or the vegetative multiplication of outstanding genotypes; the former is generally cheaper and the latter generally quicker. The additional cost of vegetatively propagated plants increases with the difficulty of propagation, and is typically a minimum of 30 to 50 per cent more than that of seedlings. Consequently, Burdon (1989) suggested that clonal forestry on an operational scale may be worthwhile only for those taxa which are easily propagated.

The means by which each of these activities is achieved and co-ordinated is described as the breeding strategy. A variety of simple and relatively cheap breeding strategies are available (e.g. Barnes and Mullin 1989; Namkoong 1989) and adaptable to the many circumstances of plantation forestry. There is sufficient genetic variation in, and control of most traits to allow productivity gains in the order of 30 to 50 per cent from the first generation of breeding; programmes based on short-rotation species have demonstrated that substantial gains can be maintained over at least several generations (e.g. Franklin 1989). The cost of genetically-improved material is a trivial proportion, typically not more than a few per cent, of establishment costs. Where economic analyses have been conducted, they suggest that investment in tree improvement as part of a plantation programme is well justified, especially where rotations are relatively short.

Adequate genetic resources of potentially useful species are the biological basis of plantation forestry. Efforts to enhance their *in* and *ex situ* conservation status, their availability, and their improvement are important to ensure the flexibility necessary for plantation forests to respond to changing demands and pressures.

Genetically transformed trees

The integration of developments in biotechnology into tree breeding programmes offers further opportunities for more accelerated and targeted genetic improvement. However, the use of these new, sophisticated, and expensive technologies should be seen as complementing, rather than substituting for, traditional selective breeding, which is cheap and robust by comparison and from which much genetic gain can be realized. While the application of biotechnologies to tree species will doubtless increase rapidly, the generally low unit value of most tree crops suggests that, as Cheliak and Rogers (1990) noted:

> development should be viewed as a means to enhance our abilities to solve specific problems .. successful application of biotechnology is conditional on an established and aggressive conventional [tree improvement] program. Without this, biotechnology makes little sense.

Nevertheless, the development of transgenic trees is likely to constitute part of improvement programmes in the not-too-distant future. The subject has been reviewed by Jouanin *et al.* (1993). The most rapid advances have been made with angiosperms, especially *Populus* spp. Hybrid poplars have proved particularly suitable because they are, among other things, easily propagated *in vitro*. Less has been achieved with conifers, which appear to require different approaches, though some French workers have achieved a measure of success with somatic plantlet production (Lelu *et al.* 1994a, b). A general problem is how to develop stable transgenic plants of species that are propagated by seed.

Possible traits that might be introduced into trees and have practical value include resistance to various pests and diseases, and to herbicides such as glyphosate (Shin *et al.*, 1994). Reducing, or modifying the lignin content of wood, thereby reducing the costs of delignification during pulping as well as reducing pollution, is another topic of interest (Higuchi *et al.*, 1994).

There are many problems, including ethical ones, still to be overcome including the risks of transmitting foreign DNA into wild populations: one suggested solution to this has been to develop sterile transgenic plants.

Matching species to site

In many places today there is considerable past experience of which species will grow and how they will perform, either from former crops of the same tree on the site or from nearby similar areas. Many publications have been written to guide those who have to select species for planting in particular countries or regions, for example Savill (1991) for the United Kingdom; and Bastien and Demarcq (1994) for the Massif Central of France. If experience is lacking, species trials may sometimes give unexpected results. Lines (1984), for example, showed that in the relatively polluted southern Pennine Hills of England, *Picea sitchensis* grows poorly where an experienced silviculturist would have been sure of good growth on apparently similar sites away from industrial pollution.

Exceptional climatic events can reveal inadequacies in species selection that may not be apparent in more normal periods. This is one reason why extreme caution should be observed before a relatively untried, but promising, exotic is planted on a wide scale. For example the cold winter of 1981/82 in Britain killed much *Pinus radiata*, *Cupressus macrocarpa*, and some other species previously thought to be relatively safe on certain sites. It also eliminated many *Eucalyptus* spp. that were being tried experimentally, enabling future work to be concentrated upon the hardy ones (Evans 1983). Similarly, the prolonged drought in western Europe during the summer of 1976 caused a great deal of mortality in *Fagus sylvatica* and *Larix* that had been planted on unsuitable soils.

On regional scales, guides for selecting species based on soils alone have proved useful. For example Evans (1984) provided a method for selecting broadleaved species using the potential of different soils according to pH, texture, rooting depth, drainage, and fertility. On a broader scale, computer-based aids for matching species to regions and sites have been developed for some parts of the world, especially developing tropical regions (e.g. Booth 1995). These are based on soil and climatic data, including rainfall and extremes and means of temperature.

The use of indicator plants in the ground vegetation

The composition of natural or semi-natural ground vegetation is often used to indicate soil and climate within restricted geographic areas. It can be a valuable guide for species selection since changes in the natural vegetation reflect and integrate changes in the physical environment. Both tree growth and the composition of the lower vegetation are, to a large extent, determined by the same basic variables of temperature, light, water supply, soil aeration, and fertility. All of these are quite difficult to measure in practice. Where there is enough natural or semi-natural vegetation, it can be a remarkably sensitive indicator of site variability in areas of uniform climate, soil, and land surface (Kilian 1981). Classifications based on vegetation are commonly used in Europe and elsewhere. For example the main types of ground vegetation in the British Isles have been widely used as indicators for species selection since the time of Anderson's (1961) work on the subject. The basis of all systems is that various indicator plants, or plant communities, are used to give a guide to the productive potential of the site (see other examples in Chapter 5, p. 65), and other features of interest. In many areas, and especially in hilly country, a detailed consideration of local topography and sites may suggest a change of species quite frequently to take account of local conditions.

Satisfying objectives of the planting scheme

The suitability of a species in economic or other terms for the objective of a plantation scheme is obviously important, but must always be secondary to the basic biological considerations of what will grow well in the soil and climate of the area. Within these limits, desirable features include ease of establishment, rapid early growth, wind firmness, and economic value of produce. For commercial plantations, the rotation length and discount rate applied can have strong bearings on choice of species. This often means a choice between quantity and quality (Price 1989). If discount rates are high, fast-growing, short-rotation species like *Pinus*, *Picea*, and *Eucalyptus*, which supply bulk markets, usually give much better returns. They therefore appear more attractive than slower-growing species, even though the latter may be more valuable per cubic metre. Market risks also influence choice: traditionally accepted species for which there are a wide range of uses are usually considered safe and are therefore more widely planted.

On most sites there is usually a relatively wide range of species available from which to choose, though in practice it is normal to concentrate on very few to avoid management and marketing problems. In most temperate regions, including Europe, plantations are established primarily for industrial purposes such as the production of saw timber, fibre, and panel products. Some are planted for environmental protection of various kinds—erosion control, windbreaks, or rehabilitation of degraded land—others for amenity, sport, or conservation. Many plantations have multiple objectives and aim to supply more than one market or purpose. Insofar as generalizations are possible, most plantations in warmer temperate countries are established for fibre, for example the extensive *Eucalyptus* plantations in Portugal. Those in the colder regions, where rotations are longer, are grown for saw timber, plywood, and veneer for which *Pinus*, *Picea*, and *Pseudotsuga menziesii* are very suitable. Much of western Europe falls between these two extremes.

7 Plantation establishment and restocking

After species selection, the next process in the management of a plantation is how to establish the new crop. Establishment is most commonly carried out by planting young trees which have been raised in nurseries or, more rarely, by direct seeding of the site.

Planting stock

Plants used in establishment programmes may be raised in several ways:

Bare-rooted plants
  Seedlings
  Transplants   } Raised from seed
Container plants

  Cuttings/setts
  Tissue culture   } Vegetative propagation

The aim is to produce plants which are cheap and have qualities and dimensions relating to handling, survival, and growth that make them suitable for use in the forest. Plants must be robust enough not to suffer damage when handled, not desiccate too quickly, have adequate reserves to make new growth, and have root systems which will rapidly become established.

In most temperate countries the techniques for growing plants by any of these methods are now so specialized that they are the concern of nursery managers rather than foresters. However, it is important for foresters to understand the qualities and potential problems associated with different kinds of plants and to specify clearly their requirements when ordering.

Bare-rooted plants

The techniques for raising bare-rooted plants are described in detail by Aldhous and Mason (1994) among others. Essentially, the nursery cycle involves sowing seedbeds in the spring and then a year or more later the seedlings are transplanted into beds where they have more room to develop. Alternatively, seedlings may be undercut at a depth of about 10 cm to promote fibrous rooting and other conditioning responses. The young plants are moved to their final position in the forest between 1 and 5 years after sowing the seed, depending on species, local climate, and the size of plants required.

An important aim of transplanting or undercutting is to encourage the formation of a root system which will ensure good survival and growth after planting. The root system should not be unduly spread, and it must be possible to lift it reasonably intact. The ratio of root weight to shoot weight is inevitably influenced by transplanting and is often lower than in undisturbed seedlings of the same age, being about 1:2 at the time of planting compared to at least 1:1 (and probably 3:1) in natural seedlings. Root length to shoot length, ratios of 1:1 to 1:2 are common in nursery plants, whether bare-rooted or container-grown, whereas 4:1 is more common in natural seedlings of the same age (Stein 1978).

Much work has been devoted to determining the nursery treatments which ensure good survival and growth after planting. These include spacing, fertilizing, and undercutting (see Driessche 1980, 1982, 1983, 1984). Many morphological standards of quality exist for bare-rooted plants (for example Chavasse 1979; Aldhous 1989; Aldhous and Mason 1994; Kramer and Spellman 1980) of which the most useful indicator is root collar diameter. However, there is often a poor relationship between seedling grade and survival. This is at least partly because during the period between lifting from the nursery and planting all the efforts and care taken to produce good-quality stock can be undone by neglect, resulting in desiccation or physical damage during storage or transit (Tabbush 1987). The importance of care in these respects cannot be overemphasized.

Physiological standards of quality and recognition of preconditioning are both important. Currently, two measures of vitality are in common use: root growth potential and fine-root electrolyte leakage (McKay *et al.* 1994). Root growth potential is measured by stimulating root growth under controlled conditions and observing the result. Evaluation of electrolyte leakage involves measuring the electrical conductivity of a volume of distilled water in which seedling roots have been immersed for a fixed period. Damaged (unhealthy) roots will leak electrolytes more than healthy roots. In Britain, the Forestry Commission offers a plant vitality testing service for conifer seedlings and transplants based on electrolyte leakage.

Container plants

Since the 1960s, there has been a trend towards raising trees in various sorts of containers in greenhouses. When the containers are together in a batch like a honeycomb they may be referred to as cell-grown stock (Fig. 7.1). The initial development, was undertaken in the testing climates of Canada and Scandinavia where it can take 4 or 5 years to grow usable transplants in conventional nurseries compared with a year or less in greenhouses. The production of container plants has been a long established practice in the tropics. There are many attractions of using container-grown stock:

1. It is possible to extend the planting season beyond the normal limits for bare-rooted stock. This is of value to management if labour is scarce.

2. The short production period, of a few months to a year, makes a good match possible between plant requirements and plant production.

3. A high output rate per person is possible, with container plant production leading to lower supervisory costs than in conventional nurseries.

4. More uniform planting stock often results (Jinks 1994).

5. Container plants usually suffer less from neglect and bad handling than do bare-rooted plants, though there is a danger of their being treated more roughly as a result.

6. If done properly, planting can be achieved in a highly consistent manner. There may therefore be less chance of damage from bad planting techniques.

7. Depending upon the size of container, survival can be better and early growth improved. This is particularly true with difficult species such as *Pinus nigra* subsp. *laricio* on drought-prone sites.

Fig. 7.1 Cell-grown pine and spruce seedlings in Sweden. After delivery to the planting site and before planting, daily watering is essential for plant survival.

However, container plants are not without their disadvantages. The main one is that to compete in cost with bare-rooted plants, containers must be small so that handling and transport costs are reduced. This means that the plants are also small, often much smaller than conventional transplants. They can therefore be more susceptible to various sorts of damage, particularly from browsing animals and birds, frost heaving, and suppression by weeds. The establishment period can also be longer. The roots of plants grown in some types of containers become deformed. Much has been done to improve container design and growing techniques to prevent this. However, bare-rooted plants commonly have root defects too (see p. 100). Finally, container seedlings are more susceptible to damage from low temperatures in the nursery because the roots, the most sensitive parts of the plants, are above ground.

Numerous studies have compared the performance of bare-rooted plants with container stock. In general, when properly planted there is little difference between the two, although if planting is delayed to late spring the root plug of the container-raised plant gives the plant a definite advantage over bare-rooted stock (Kerr and Jinks 1994; Burgess *et al*. 1996).

Vegetative propagation and tissue culture

It is normal practice to use rooted cuttings, sometimes called setts, for propagating *Populus* spp. (see p. 227), *Salix* spp., some *Ulmus* spp., and a few other species which are easy to root and where clonal material exhibiting disease resistance or exceptional vigour is important (Heybroek 1981). Cuttings are also used for propagating trees which do not produce viable seed, such as Leyland cypress which is an intergeneric hybrid between *Chamaecyparis nootkatensis* and *Cupressus macrocarpa*, and occasional interesting cultivars of many species. They are also used to establish seed orchards of selected trees, such as *Picea sitchensis* in the United Kingdom. The use of mass propagation of cuttings of the more common plantation conifers is now widespread as a means of rapidly introducing improved genetic stock (Mason 1992). Gains in productivity of at least 10 per cent, and possibly much more, are expected compared with stock from seed orchard origin. The cost of plants is between 20 and 50 per cent greater than conventional seedlings or transplants, depending upon the age and size of plant required.

Mason and Jinks (1994) have discussed many of the techniques and problems associated with vegetative propagation. Success at rooting is critically influenced by the physical conditions that the cuttings are subjected to. Substrate, nutrients, humidity, temperature, light, and application of auxins are particularly important as well as physiological factors, including particularly the age of the tree from which the cutting is taken, its position on the crown, and its general health. Generally material which is less than 6 years old from seed is the easiest to propagate.

Tissue culture has even more potential for vegetative propagation through extraordinarily high rates of multiplication. The term is used to describe three areas of vegetative propagation:

1. True tissue culture, which involves formation of an undifferentiated callus before subsequent organ formation.

2. Organ culture, where organized regions of cells, usually bud meristems, are cultured but remain physiologically intact.

3. Cell suspensions where the unit to be cultured is a single protoplast (that is, a plant cell with the cell wall removed).

Techniques have advanced considerably and many tree species are successfully replicated in this way, for example *Prunus avium* (Label *et al.* 1989), *Picea sitchensis*, and *Larix kaempferi*. Tissue culture and micropropagation (Fig. 7.2) lend themselves to genetic engineering—see Chapter 6.

Being able to raise clonal material in quantity enables the rapid replacement of forest crops by selected, genetically superior trees in terms of vigour, form, wood properties, disease resistance, and other attributes which are thought desirable. Superiority is achieved by dominance or epistasis which may break down in sexual propagation. Clonal propagation has been practised for centuries with apples, citrus fruits, and rubber. These are relatively valuable crops on which expensive protection measures can be justified: timber crops are not. A disadvantage of single

Fig. 7.2 Micropropagation—a young shoot growing from an embryonic axis of a *Juglans nigra* × *J. regia* hybrid after 6 weeks of *in vitro* culture. (Photo: C. Jay-Allemand, INRA.)

clones is that, if widely planted, they are likely to become exceptionally vulnerable to diseases and pests which are able to overcome any temporary resistance bound to one or a few genes. This has occurred in Japan with *Cryptomeria japonica*, and with *Populus* spp. in Europe. The problems associated with a narrow genetic base are well known in various agricultural crops. The risk may be reduced in a plantation by mixing numerous selected clones (see p. 28). Vegetative propagation is seen by many as useful for studies in forest genetics but it is not a technique to be used on an extensive scale unless clones have undergone a long (two rotation) period of successful growth. The problems and possible advantages of clonal forestry are still being debated.

Direct seeding

Direct seeding can complement conventional planting and natural regeneration when appropriate to the objects of management, site, and species. Since repellents were developed for protecting seeds from predators the technique has been used with success in several temperate countries. For example about 10 per cent of afforestation in France is done by seeding (Kroth *et al.* 1976), and rather less than 5 per cent in Scandinavia. Its main value in temperate countries has been in extensively managed areas, on difficult terrain, after large fires, in windthrown areas, and in forests devastated by insects. Seed is often broadcast from the air. Species appropriate for direct seeding must normally have small and easily available seeds which germinate rapidly on the soil surface and then grow quickly. Trials in Britain of sowing *Quercus* and *Fraxinus* have been of variable success. Control of competing weeds and prevention of browsing are essential.

Root form of planted versus naturally seeded trees

The debate about the extent to which planted trees suffer from root deformations has been renewed every 20 to 30 years since the 1880s (Huuri 1978).

Naturally regenerated and direct seeded trees normally develop an array of sturdy lateral roots radiating from a well-developed tap root. The tap root grows downwards to an extent which depends upon the nature of the soil and the species. Some species, and especially *Pinus*, lose the capacity to initiate first-order lateral roots (which eventually become the main structural elements) early in their first season of growth (Burdett 1978, 1979), possibly within 60 days of sowing seed. Thus, the final configuration of a root system may be established very early in life and remain essentially unchanged. Other species such as *Picea abies* are much more adaptable.

The number of roots, their growth, and branching habit control the shape, size, and symmetry of natural root systems of trees. In *Picea* and some other species adapted to wetter soils, lateral and oblique roots take over the function of structural support as the growth of the taproot slows. The root system takes on a typically bell-shaped form, penetrating as far as soil conditions permit. In some pines and other species like *Pseudotsuga menziesii* and *Quercus* the taproot is relatively more important. Natural root systems are normally effective anchorages for trees, as attested by the wind firmness and upright posture of most seeded trees.

Though asymmetry of the root system is sometimes a natural feature, particularly of non-dominant trees and in trees on stony and sloping sites (Eis 1978), the roots of many planted trees are generally much less adapted to the function of mechanical support at the outset. Root systems of bare-rooted plants are modified by pruning, lifting, and transplanting. The roots of container plants are also shaped and pruned. All roots may be distorted on planting. Bare-rooted trees, even if well planted, are likely to have bilaterally compressed root systems. If trees are not properly planted, all roots can be unilaterally distributed in relation to the base of the stem, or they may be squashed into a planting hole that is too small. Container-grown stock also has characteristic modifications. In some containers major lateral roots tend to lie parallel to the taproot for the first 10 to 15 cm of their length instead of growing in an approximately horizontal plane. Sometimes roots grow around the container walls in a spiral fashion, giving rise to distortions and instability of the planted tree.

There is general agreement that natural regeneration or direct seeding offer the best prospects for good root development. There is little hard evidence that planted trees, whether of bare-rooted, container-grown, or cutting stock, suffer seriously impaired root development as the huge areas of successfully planted forests demonstrate. Moreover, while the pattern of root growth can be considerably influenced by nursery and planting practices, the subsequent development of a root system is often more affected by soil physical conditions (such as rooting depth, impervious layers, and anaerobic conditions), by cultivation and ploughing on the planting site itself, and the proximity of the planting position to old stumps (Quine *et al.* 1995).

Growth in relation to rooting characteristics

Some studies have demonstrated that in comparable conditions planted trees (both transplants and container plants) do not grow as well initially as naturally regenerated material of the same age. Bare-rooted plants of most species may 'check' during the year of planting, especially where there is much competing vegetation (Hellum 1978; Leaf *et al.* 1978). *Picea* spp. are particularly prone to this check. Container-grown plants may display a lack of root growth compared with top growth, and can have shoot to root ratios of over 25 to 30 per cent higher than natural seedlings of the same age after 10 years. This may be due to the difficulty roots have in some soils in moving from the medium of the container into the surrounding soil.

By contrast naturally sown or directly seeded plants can struggle to grow in competition with dense ground vegetation. The healthy, sturdy transplant can show superior survival and vigour in such situations.

Size of plants

Variations in plant size in the seedbeds and transplant lines of forest nurseries, and in container plants, are usually obvious. It is desirable to know to what extent this diversity indicates inherent differences in vigour and how size might affect survival during the first few years after planting.

Research in the 1960s by Sweet and Waring (1966) with *Larix kaempferi* and *Betula* suggested that early size variations found in nurseries arise from different rates of germination and variations in seed size as well as from genetic differences in vigour. They concluded that much of the normal size variation originates at the time of germination and is, at least partly, non-genetic in nature.

Pawsey (1972) showed that, within fairly wide limits, survival and subsequent growth in the field of different-sized plants of *Pinus radiata* is essentially the same if the various size classes are planted separately from each other. However, where large and small seedlings are mixed together, the superior growth of larger plants can be maintained indefinitely by virtue of their original slight size advantage. Pawsey concluded that seedling size is not a sufficiently reliable indicator of inherent vigour to warrant the rejection of small-sized plants. However, in practice, mortality is often much higher in small plants owing to their delicate nature, risk of desiccation, and susceptibility to weed competition.

Among conifers, bare-rooted transplants 30 to 40 cm tall and with proportionally thick root-collar diameters are usually favoured but there are exceptions to this general rule on size, especially with some pines. For example the root system of *Pinus nigra* subsp. *laricio* is particularly sensitive to damage during lifting from the nursery and transplants can suffer very heavy losses after planting. For this reason small-sized container plants are often used even though they may require more weeding. Large plants can be more prone to basal sweep and early instability than small ones.

With many broadleaved species, rather taller plants can be successfully used, and these often range up to 1 m in height, but sturdy transplants with at least a 5 mm root-collar diameter are usually the best choice. Container stock is normally 15 to 30 cm tall, depending on container size. However, for both bare-rooted and container stock, good survival after planting is strongly correlated with root-collar diameter and, as such, it is a more important criterion than height.

Individual tree shelters

The establishment of broadleaved species is significantly helped by the use of individual tree shelters (Tuley 1983; Potter 1991; Kerr and Evans 1993). These are plastic tubes with cross-sectional areas of about 80 cm² which enclose recently planted trees (Fig. 7.3). During the critical first 3 years in shelters, height growth can be accelerated by three times in *Quercus* and several other broadleaved species: conifers are less responsive. This rapid growth reduces the period of susceptibility to weeds and frost damage. The shelters also give protection from small grazing mammals such as roe deer and rabbits, and enable chemical herbicides to be used more safely. On very weedy sites it is also possible to see clearly where the planted trees are. Tree shelters 1.2 m tall are commonest and give protection against browsing animals as big as roe deer; taller shelters are needed if fallow or red deer are present and smaller than 1.2 m if none of these deer species is present. Although modern tree shelters will break down in sunlight, trees planted in older types often need to be released, to avoid strangulation. Many millions of tree shelters are used in Europe each year.

Replacing dead plants

The replacement of dead seedlings is often carried out as a routine measure after planting. It can be a very wasteful operation. The original crop has at least a year's start on the replacements which can fall far behind in growth with only a remote chance of being included in the final crop. Only at very wide spacings do replacements have a significant effect on the original crop in terms of height or

Fig. 7.3 New planting on farmland. Tree shelters here are used for growing oak. (Crown Copyright.) Areas larger than about 1 ha should normally be fenced rather than using individual tree shelters.

diameter growth or branch size. Unless failures are numerous, or planting very wide, or gaps very large, larger than the space occupied by one final crop tree (about 25 m^2), replacement is not usually considered worthwhile. If it is necessary, then the plants used should be especially large and vigorous so that they are able to compete with the original trees. The best strategy is to ensure that the original planting is completely successful. If many failures do occur, say more than 20 per cent of the original planting, the reasons should be investigated and remedies found, rather than simply planting more trees. The causes of high mortality might be browsing, poor weed control, or dead or unhealthy plants coming from the nursery.

Planting methods

Even though the way in which trees are planted is known to affect future rooting patterns, growth, and stability, planting is seldom given the care it deserves. Tree planters are often paid according to how many trees are planted per day, not according to how well they are planted. It is well established, for example, that the growth of a tree after planting is positively correlated with the volume of soil disturbed. Mullin (1974) investigated how planting by two different squads affected the growth of red pine (*Pinus resinosa*). After 20 years one squad's planting produced a stand with 14 per cent more volume than the other's, indicating the possible long-term effect of bad planting.

Many methods are used for planting bare-rooted trees and these often depend upon personal preference and experience. The simplest, and most common, is slit notching in which a single vertical cut is made with a spade or mattock (Fig. 7.4). The spade is then pushed backwards and forwards to break up the soil at the bottom of the notch and the plant inserted with the roots 'as well spread as possible' (Hibberd 1985). This inevitably leads to a bilaterally compressed root system, at best. The soil is then pushed back into position and firmed around the root collar without excessive compaction. Variants to this are 'T' or 'L' notches where a second dig of the spade is done at right angles. On organic soils 'plugs' of peat are sometimes removed with semi-circular planting spades and then replaced after inserting the plant. On suitable terrain (usually relatively flat, stone- and stump-free sites), mechanical planting with modified agricultural 'cabbage planters' and other machines is sometimes practised.

The process of tree planting clearly introduces a measure of localized cultivation to the soil. It can vary from no more than opening a slit in the soil to digging a large pit in which the tree's roots can be spread. The much greater cultivation of the latter usually leads to better survival and better initial growth, but these improvements are rarely considered justifiable for forest scale planting because of the much greater costs involved. An experienced worker can, for example, plant about 200 trees a day by pit planting and up to 1000 by notching.

Fig. 7.4 Planting trees in an upland forest in Britain. (Crown Copyright.)

Position of planting

Planting positions are usually varied according to the method of ground preparation, expected competition from weeds, moisture regime, and exposure. On
drought-prone sites, planting is often done in a ploughed furrow so that the plants
can benefit from the weed-free conditions and moisture at lower levels. On wet
ground, by contrast, trees may be planted on top of, or on a step in the side of, a
mound or an inverted ridge, to benefit from drainage. Steps are used when the
ridge is very thick or the site severely exposed. Where conditions are less extreme
plants are inserted directly into the normal unprepared soil surface. Except on wet
sites where mounding is usual, it is normally considered too costly to carry out
any form of ground preparation on most second-rotation (restocking) sites in
Britain, because of old stumps, though in Scandinavia ground preparation is
normal almost everywhere before replanting.

Time of planting

The prospects for successful establishment are greatly improved if bare-rooted
stock is planted when it is dormant, although planting must not be attempted when
there is deep snow or the ground is frozen. Thus, most planting is carried out in
late autumn or early to mid-spring, before bud burst. Cold storage can extend the

last planting date by a few weeks provided conditions in late spring/early summer are still favourable (cool and moist). Container plants can be planted throughout the growing season but best results are still obtained in the autumn or spring (Burgess *et al.* 1996; Kerr and Jinks 1994).

With broadleaved and coniferous bare-rooted plants, opinion is divided as to whether planting should be carried out in the autumn and early winter, or in later winter and early spring. Sometimes there may be no choice because of the size of the planting programme and the labour force available to carry it out. Local preferences are often unconscious expressions of anticipated peaks of root growth potential (see next paragraph) for the area and species. This is particularly the case in Europe. In Britain, spring planting tends to be favoured for conifers in upland forests, and for broadleaves in the lowlands autumn planting is generally preferred. In climates with a prolonged dry season or where winter snow and frost are normal, the planting season may be very short if success is to be assured, and it will vary according to the particular climatic conditions.

Survival is critically influenced by the root growth potential of newly planted trees; this is the ability to initiate or elongate new roots shortly after planting out (Ritchie and Dunlap 1980). Typically this potential increases during the autumn and winter months, peaks in late winter or early spring, and declines rapidly just prior to vegetative bud burst. There may sometimes be a minor increase in mid or late summer. It is also lower on sites suffering from drought. Survival is illustrated for one particular year on an upland site in Northern Ireland, when March and April planting gave the best results (Fig. 7.5). Though this is commonly the case, very dry, windy weather in early spring can result in high mortality.

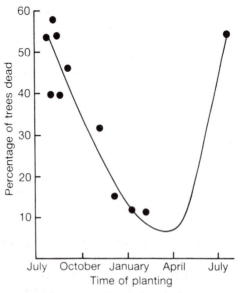

Fig. 7.5 Effect of the time of planting on mortality of Sitka spruce (*Picea sitchensis*) planted on deep peat in Northern Ireland. (Northern Ireland Forest Service, unpublished.)

Pattern and stocking of planting

Plantations are almost always established using regular spacings to ensure a reasonably uniform development of the crop. This involves planting in lines which can easily be followed, at least in one direction. It is common to plant in a rectangular rather than square pattern, with the wider spacing of 2.2 m or more between rows to facilitate the passage of tractor-borne implements and to make operations such as chemical weed control cheaper.

The actual density adopted varies according to species, sites, and management intentions. It aims at the best compromise between establishment costs, the likely costs of future operations, and the expected value of produce. There are no definite rules but commonly close spacings are used to eliminate competing vegetation quickly, reduce fire risk, and suppress side branches and improve upward development in many broadleaved species (Kerr and Evans 1993). Wider spacings are more appropriate for fast-growing species, especially in areas which will never be thinned or where early thinnings are unprofitable. However, wide spacing which leads, in conifers, to excessive juvenile wood development and in broadleaves to poorly formed heavily branched trees, can prove a false economy. These aspects are discussed further in Chapter 10.

Planting mixtures of species and 'nurses'

Pure, even-aged crops tend to predominate everywhere that plantation forestry is practised due to their financial advantages. Management, work in the forest, harvesting, and marketing are all simpler. Mixed crops, and especially uneven-aged mixtures, have certain biological, biodiversity, and aesthetic advantages, besides a few financial ones. For example mixed stands may be more flexible in meeting market demands and mistakes with species selection are not so serious. However, the most common reasons for planting mixtures are for amenity and nature conservation, for providing early returns from a conifer when a final crop of broadleaved species is intended, and for the 'nursing' effect of one species on another.

Nurses perform one or more of three functions. The first is to suppress competing vegetation rapidly. The second is to protect tender species from frost and exposure during their early years when they are particularly sensitive. Late successional species such as *Fagus sylvatica*, several *Abies* spp., and *Thuja plicata* may all grow better initially if mixed with a faster growing nurse or if planted under the canopy of a near-mature crop (Fig. 7.6). *Quercus*, also, tends to develop a more pronounced and straight central axis if grown in mixture with conifers.

The third use of nurses is on impoverished sites where one species may benefit the nutrition of another. Nitrogen-fixing species, whether trees, such as many *Alnus* spp., shrubs, or even herbs are often used as nurses, but other trees can also benefit nutrition, see examples in Moffat and McNeill (1994). For a specific example, O'Carroll

Fig. 7.6   Beech (*Fagus sylvatica*), a strongly shade-bearing species, being established together with Douglas fir (*Pseudotsuga menziesii*) under a canopy of *Betula pendula* at Arundel forest, England. (Photo: D. Cooper.)

(1978) found that on a peaty podzol in Ireland, *Picea sitchensis*, intimately mixed with *Larix kaempferi* or *Pinus contorta*, grew significantly better than *Picea* alone due largely to improved nitrogen nutrition caused especially by *Larix* (Table 7.1).

How this improved nutrition is brought about is not clear but it may be by mobilization and rapid turnover of nitrogen by the deciduous *Larix* or improved mycor-

Table 7.1 Effect of Japanese larch (*Larix kaempferi*) and lodgepole pine (*Pinus contorta*) on growth and foliar nutrient contents of Sitka spruce (*Picea sitchensis*) (from O'Carroll 1978)

Treatment	Mean height at age 18 (cm)	Height growth in year 18 (cm)	Foliar nutrient concentration (% dry matter)		
			N	P	K
Pure Sitka spruce	119	10	1.11	0.16	1.08
Spruce/larch	295	34	1.58	0.21	1.24
Spruce/pine	151	16	1.42	0.17	1.08
Least significant difference (LSD) 5%	23***	12***	0.23***	0.03***	NS

*** indicates significance at the 99.9% probability level.

rhizal associations. Similar effects have been reported on peat sites in Scotland by McIntosh (1983), and Evans (1984) gives several examples of the growth of broadleaved trees being improved by *Larix kaempferi* and *Pinus sylvestris*.

'Self-thinning' mixtures are used on sites of high windthrow risk where thinning operations may be too dangerous (Lines 1996). These can be achieved by planting mixtures of species with different growth rates such as the Queen Charlotte Islands provenance of *Picea sitchensis* with Alaskan *Pinus contorta*, or by using fast- and slow-growing provenances of the same species, such as Queen Charlotte Islands (fast) and Alaskan (slow) *Picea sitchensis*. Both of these mixtures are used in Britain. The slow-growing trees might limit the branch size and size of the core of juvenile wood of the faster ones before being suppressed, leaving the stand effectively at a wider spacing and capable of producing sawlogs rather than pulpwood.

Mixtures are difficult to manage because the more valuable or long-term species is often slower growing. If thinning and liberation is delayed, this species can easily be suppressed and lost.

Planted mixtures are commonly planted in alternating rows of the different species. In southern Britain, one or two rows of *Quercus* are usually planted between every three, four, or five rows of *Picea abies*. More rarely *Quercus* is planted in groups of at least nine trees, within a matrix of the conifers. Such practices should be avoided on hillsides where they can result in unsightly striped or chequer-board effects. The important principle is, however, to ensure robust mixtures, that is

Fig. 7.7 A planted mixture of ash (*Fraxinus excelsior*) and western red cedar (*Thuja plicata*). Originally the mixture was three rows of red cedar alternating with one of ash, but thinning has now reduced this to alternate rows of each species. These two species have similar early growth rates.

mixtures in which a small amount of neglect will not irreparably harm the long-term potential of the final crop, usually slower-growing, species. This is achieved by ensuring that the rates of growth in height of the two species do not differ too much: the faster growing species should not exceed the slower by more than 20 per cent, otherwise there is risk of suppression (Fig. 7.7).

Some broadleaved species, most notably *Quercus* when grown for high quality timber, require fairly dense stand conditions to ensure the development of a reasonably formed stem and upward growth (Kerr and Evans 1993). If, because of cost, *Quercus* is to be planted at a wide spacing (more than 3 m apart), then mixtures may be appropriate: either planting it with carefully chosen conifers, or retaining natural regeneration of, for example, *Betula*, *Alnus*, or *Salix* spp..

Conclusion

There are many components to the successful establishment of tree plantations. To be successful, particular attention must be paid to—the quality of the plant, selection of the site and its preparation, the species chosen, skill in planting, aftercare (weed control), and protection. Neglect of any one can lead to failure of the whole project.

8 Weed control and cleaning

During the early stages of establishment of a tree crop, whether it originates from a plantation or from natural regeneration, the high availability of light, water, and nutrients can lead to an explosive development of herbaceous or woody vegetation. A plant is usually regarded as a weed if it grows where it is not wanted and is a positive nuisance in some way. However, some plants can have beneficial effects on the desired tree crop. Management of forest vegetation consists of selecting plants which are considered as favourable and eliminating those considered undesirable (Gama *et al.* 1987).

This chapter first describes the different ways in which forest vegetation may interact with the trees and the main characteristics of weeds, then gives indications of appropriate strategies for controlling them. Particular emphasis is put on the use of herbicides, due to their high efficiency and low cost, and their importance in agriculture and forestry.

Forest vegetation interactions

Competition

Forest vegetation can compete with young trees for light, water, and nutrients and hence retard growth. Competition for water appears to be the most important factor (Lévy *et al.* 1990). In areas which suffer from drought, or even mild deficits of summer water, weed control may be critical for the survival of young trees and it can result in dramatic improvements to growth (Fig. 8.1). During a particularly dry year in central France, Frochot (1988) showed that the survival of a *Pinus nigra* subsp. *laricio* plantation was 61 per cent in weeded plots compared to 27 per cent in control plots. Other examples are illustrated in Figs 8.2 and 8.3.

Other negative effects of weeds

Where resources are not so limiting, it may still be necessary to control weeds to prevent them smothering young trees. Climbers or creepers can lead to permanent deformations or even strangulation of the trees. Wounds can appear where unwanted woody species rub against the crop trees. Some plants such as mistletoe

Fig. 8.1 *Pinus pinaster* in the Landes, France. Disc cultivation for weed control has been carried out between lines to reduce competition for water and to reduce the fire hazard.

Fig. 8.2 The effects of competition with grasses on 7-year-old *Pinus sylvestris*—no control (left), and complete control (right) on a 1 m diameter spot round the tree. (Photo: L. Wehrlen, INRA.)

Fig. 8.3 Twelve-year-old *Picea abies* planted on a heather (*Calluna vulgaris*) covered site 2 years after the heather round the taller trees (right) had been treated with 2,4-D. No heather control was carried out on the left. (Photo: L. Wehrlen, INRA.)

are parasites and others can provide habitats for predators of organisms which damage crop plants. Fryer and Makepeace (1977) showed how plants may indirectly affect tree growth, by being primary or secondary hosts for pests or parasites: for example species of *Ribes* are alternate hosts for the fungus *Cronartium ribicola* which causes 'blister rusts' on five-needled soft pines (see p. 168). Weeds also provide cover for rodents and a reduction in cover can prevent damaging increases in their populations.

Some weeds are also allelopathic in that they form chemical compounds that are toxic to, or inhibit, potential competitors. Baker (1974) states that allelopathy of this sort seems to be proven in a growing number of cases. It can affect germination or growth or it can indirectly affect nitrogen nutrition by interaction with mycorrhizas (Gama *et al.* 1987). It is, however, difficult to demonstrate the real impact of allelopathy in forest conditions due to the competition for water, light, or nutrients (Frochot *et al.* 1990).

In areas prone to a severe fire risk, the control of all ground vegetation can become a necessity, if not on the entire forest area, at least on firebreaks (p. 199).

Accompanying vegetation

Forest vegetation is not wholly detrimental. It can offer protection against a number of physical risks, such as late frosts, wind, or sun scorch. It can, in some cases, assist natural pruning and can have interactions—good or bad—with game. Ground vegetation in forests can be useful in protecting soil structure, preventing erosion, and acting as a temporary 'reservoir' for nutrients in young crops—nutrients which might otherwise be lost from the site. In the sense that many woodland plants may be part of a natural or semi-natural community of organisms, any reduction in the diversity of vegetation will affect the nature conservation value of the site.

Species are not necessarily weeds throughout the whole of their ranges or the whole of their lifetime. Bracken (*Pteridium aquilinum*) is ubiquitous, yet is only recorded as a serious forest weed on acid soils. *Rhododendron ponticum* becomes particularly aggressive and very difficult to deal with in the wetter and milder western parts of Britain (Tabbush and Williamson 1987) and Ireland where it is one of the few exotic forest weeds. Nor are weeds necessarily considered serious problems under all silvicultural systems within a particular area. Heterogeneous systems in both agriculture and forestry are often characterized by a lack of weeds, as well as pests and diseases. Clear-felling and replanting usually cause the worst difficulties.

Combinations of positive and negative effects are very often encountered. Frochot and Trichet (1988) have shown that small weed-free areas first affect young *Pinus sylvestris* growth beneficially in comparison with larger weed-free spots, due to the sheltering effect of the surrounding vegetation. Small weed-free spots are detrimental, however, in later years due to competition for water. Unweeded controls were always smaller.

Characteristics of weeds

An 'ideal' weed has a number of the special features listed in Table 8.1. Many are characteristic of *r* -selected species (p. 165). Fortunately no plants have all these features but those that have none, or very few of them, are unlikely to be weeds. These characteristics lead to the problems which cause plants to be called weeds: they establish themselves without deliberate action by man; they are difficult to control because they are competitive, adaptable, and tend to form extensive populations.

The main detrimental effect of weeds is to reduce tree survival after planting (Frochot 1988). When disturbance first comes to an area of natural vegetation such as woodland the weeds that appear are often species which are native to the area, whereas agricultural weed floras may contain a substantial proportion of non-native species as well. Weeds are nearly always present, at least in the early stages, in the life of a plantation.

The major competitive plants in Europe are the grasses (Dohrenbusch and Frochot 1993) *Deschampsia flexuosa, D. cespitosa, Holcus mollis, H. lanatus, Brachypodium* spp., *Agrostis* spp., *Molinia caerulea, Elymus repens, Festuca*

Table 8.1 Ideal weed characteristics (adapted from Baker 1974)

(1) Germination requirements fulfilled in many environments.

(2) Discontinuous germination (internally controlled) and great longevity of seed.

(3) Rapid growth through the vegetative phase to flowering.

(4) Continuous seed production for as long as growing conditions permit.

(5) Self-compatible but not completely autogenous or apomictic, though ecological races often found.

(6) When cross-pollinated, unspecialized visitors or wind used.

(7) Very high seed output in favourable environmental circumstances.

(8) Some seed produced in a wide range of environmental conditions, tolerant and plastic.

(9) Adaptations for short- and long-distance dispersal.

(10) If perennial, vigorous vegetative reproduction, or coppicing ability, or regeneration from fragments, and mechanisms that permit survival under temporarily unfavourable conditions, e.g. altering growth and development in response to changes in environment.

(11) If perennial, brittle, so not easily pulled from the ground, or unpalatable to grazing animals.

(12) Ability to compete interspecifically by special means (rosette, choking growth, allelopathic chemicals).

gigantea (see Fig. 8.2) and the sedge *Carex brizoides*. In eastern Europe, *Calamagrostis epigeijos* is expanding due to the widespread use of nitrogen fertilizers. Other plant groups are also of importance: *Clematis vitalba* and *Lonicera periclymenum* and shrubs such as *Cytisus scoparius*, *Calluna vulgaris* (Fig. 8.3), and *Ulex europaeus* have expanded in western Europe. Trees such as *Betula, Salix,* and *Carpinus betulus* are sometimes considered weeds, and *Rubus fruticosus* is a serious problem in many European forests. Typical Mediterranean woody shrubs such as *Rosmarinus, Thymus,* and *Lavandula* can also become problems in southern European forests.

Strategies of weed control

In forest plantations, crop trees suppress most competing herbaceous and many woody weeds once the crowns begin to touch. Weed control is normally only important before this stage. Unlike annual crops, if the growth of a forest plantation is somewhat slowed for a few years by competition with weeds, the economic losses are likely to be relatively small, provided that survival has not been badly affected. After the initial establishment period, crop trees at the thicket stage may sometimes need liberating from unwanted perennial weeds, often naturally seeding species of trees such as *Betula*, and from climbers and coppice regrowth. This is done before, or at the latest during, the first thinning. The operation is usually called 'cleaning'.

The advantages of much enhanced early growth through nearly complete weed control are rarely realized in forestry because total control is very costly. Furthermore, many weeds provide a source of food for herbivores such as deer, voles, rabbits, and sheep. In their absence, the planted trees could be severely browsed and for this reason alone complete control is not always considered desirable. Until the advent of chemical herbicides it was only possible by mulching, regular hoeing, or complete ploughing. Traditionally, complete control has only been carried out with species such as *Populus*, that are grown rapidly on short rotations, or with allelopathic weeds, such as dense *Calluna vulgaris* (Fig. 8.3) on sites planted with *Picea* (though with this problem an application of a nitrogen fertilizer may be as cost effective), and occasionally where weeds are a major fire hazard.

Five major techniques of weed control are practised in forestry and effective and rational control involves a judicious mix of those appropriate for a given situation.

1. Cultural control methods are commonly used before planting in both forestry and agriculture, usually by cultivation to produce strips of temporarily weed-free soil for planting (see Chapter 5). After planting cultural methods include:

    (a) Mulches such as peat, bark, polythene, felt, or even newspaper (Frochot and Lévy 1986) which suppress weed growth. Individual tree guards (plastic tubes) are used against wild or domestic animals, but can also be beneficial against weeds, both by mechanical protection and by the increased growth of trees which allows them to outgrow the weeds quickly (Tuley 1985; Dupraz *et al.* 1993). The tubes also offer protection to trees against the damage that can occur when practising mechanical control, or act as a shield against non-selective chemical treatments. They are, however, expensive and more commonly used in arboriculture, in agroforestry, or confined to plantings of less than 1 ha.

    (b) Hand control of weeds using grass hooks, scythes, and other cutting tools can be useful to supplement chemical or other practices and where weeds can be dealt with easily. The skill required is small, costs of equipment very low, but control is effective for short periods only. Risks to trees by cutting are slight and to operators, low.

    (c) Machine control, includes the use of portable hand-held machines such as brush cutters and clearing saws, pedestrian-controlled machines such as reciprocating cutters. Tractor-mounted machines include flails and cutting or crushing rollers. Mechanical treatments can be as cheap as chemical ones but require somewhat less skill and are safer for operators. They are effective for a short period only because they do not kill the weeds and can involve expensive equipment. Risks to the crop can be relatively important, unless done carefully.

    (d) Grazing by domestic stock can be a valuable means of control. It must be carefully managed and adapted to the age and species of trees and to the animal's behaviour: it is mostly practised when the trees are tall and large enough to support grazing pressure (p. 234).

Much effort in weeding is often devoted to grass cutting. Insley (1982) has shown that in the British climate at least, unless the grass physically smothers the trees, cutting alone does not result in significantly better growth and may even be detrimental (Davies 1984).

2. Chemical herbicides provide a cheap and effective method of weed control in plantations; however, the use of chemicals in the forest is restricted by law in most European countries. A separate section of this chapter is devoted to herbicides.

3. Fire is occasionally used to clear rank ground vegetation before planting, an example being heather burning. It is very cheap, though risky in the vicinity of plantations and requires particular skills. Prescribed burning inside plantations is practised in some European countries (France, Portugal, Scandinavia, see Chapters 5 and 13), but mostly at later stages when the trees are protected by thick bark.

4. Crop competition. Shading out of weeds by the planted trees is an important constituent in weed control. It does not often eliminate a problem but can influence the extent and duration of serious competition. It is determined by initial spacing, size of transplant, rate of growth, and choice of species among other factors. A technique termed 'competitive replacement' by Frochot *et al.* (1990) consists of sowing ground-cover plants normally used in agriculture, after site preparation and weeding with a non-persistent herbicide. An appropriate choice of seed mixture, non-competitive towards the tree crop, can delay competition from natural vegetation and help retain nutrients which will later become available for the trees. The use of nitrogen-fixing legumes may later benefit the tree crop.

5. Biological control with plant pathogens is not yet used on any significant scale, even in agriculture, but has potential. Inoculation of *Quercus* with the oak wilt fungus, *Ceratocystis fagacearum*, has been used in Minnesota to convert marginal oak stands to *Pinus*. This was cheaper and more efficient than the once conventional use of 2,4,5-T because of lower costs, easier application, greater mortality, and total lack of injury to other tree species, and safer for human use. Though little spread of the fungus occurred between treated areas and adjacent oaks, this inoculation technique met violent objections because of the possibility of oak wilt becoming epidemic (French and Schroeder 1969; Wilson 1969). Other examples of biological control are given by Templeton (1981).

The most desirable strategy of weed control is one that can eliminate or greatly reduce the need for additional measures, is to ensure rapid establishment to harness the competitive ability of the trees themselves. Cultivation is a very effective method of controlling grasses, herbs, heather (*Calluna vulgaris*), and sometimes bracken

(*Pteridium aquilinum*). Where the site is nutritionally impoverished and needs drainage, serious weed competition can be virtually eliminated by ploughing. The vegetation is buried under spaced and inverted plough ridges. By the time it begins to recover the tree crop is well on the way to occupying the site and has a lead which makes further control unnecessary.

Combinations of treatments usually prove most effective. In an experiment in southern England *Picea sitchensis*, which was kept totally weed-free and supplied with any deficient nutrients, grew at a yield class of 32 m^3 ha^{-1} year^{-1}, which is one-third more than the normal maximum of 24 m^3 ha^{-1} year^{-1} (Rollinson 1983). Frochot and Lévy (1986) showed how the combination of plastic mulch and fertilizer was most effective for promoting early height growth in *Prunus avium* in eastern France compared to other treatments involving mulch or herbicide, or an untreated control (in descending order). Frochot *et al.* (1992) has shown how the combination of black plastic mulch and individual tree protection improved height, diameter growth and tree form of young *Fraxinus excelsior*.

Chemical herbicides

On more fertile sites, especially second-rotation ones, there may be a rapid invasion of grasses, herbs, bracken (*Pteridium aquilinum*) and other ferns, and some woody weeds for which chemical herbicides are now the main counter. They are usually treated in the summer before planting. Weed tree species are seldom very numerous and are usually controlled by hand or machine cutting. More competitive weeds, including broadleaved coppice such as *Quercus* spp., *Betula* spp., or *Rhododendron ponticum* are the most troublesome. Coppice regrowth is initially far more rapid than the growth of newly planted trees. To enable the latter to survive, it can be necessary to apply herbicides to cut stumps or to control shoots by frequent herbicide and/or cutting operations. This can cost several times more than the original cost of planting. Some recommendations are described by Evans (1984). Though all the methods described have a definite place in crop establishment, expenditure on weeding has tended to decrease steadily since the mid 1960s. This is attributed by Holmes (1980) largely to the advent of chemical herbicides. Chemicals have often reduced the need for grass/herb weedings to one per season where two or three were necessary before. Similarly on sites covered with some woody weeds, only one chemical treatment may now be necessary in the life of the crop instead of many hand or machine cutting operations. A greater acceptance for environmental reasons of naturally regenerating native species among planted crops has also reduced the need for control, especially at the cleaning stage. Reductions have resulted in the establishment period caused by more effective use of fertilizers, cultivation, and more appropriate sizes of planting stock and species.

The discovery of the herbicidal properties of 2,4-D in 1944 and its subsequent widespread use heralded the rapid development of a large array of available herbicides. Chemical treatments are relatively cheap and can be very effective for a long period but require skilled operators as the risks from using inappropriate chemicals, wrong dosages, and spray drift are particular dangers to the environment. Some chemicals are also highly toxic to people and animals. The quantity of herbicide used by the forestry industry is extremely small compared to agriculture. Manufacturers have not, therefore, developed specific forestry herbicides: the existing ones have, however, been tested in forest or nursery conditions. Since the first developments of herbicides there has been a strong increase in public awareness of the dangers of misusing chemicals, particularly in forests which are considered as natural areas. This has led to important legal restrictions in most countries.

The chemical properties and mode of action of most currently used herbicides on forest vegetation, the methods of application and recommendations, and the legal aspects are described in a number of scientific publications (Du Boullay 1986; Gama *et al.* 1987; Barthod *et al.* 1990; Dohrenbusch and Frochot 1993; Willoughby and Dewar 1995), as well as information leaflets and manufacturers booklets. They are summarized briefly below (after Frochot 1990).

Herbicides can be absorbed by:

(1) leaves, through the cuticle;

(2) stems, including lignified stems, and through wounds or even bark;

(3) roots.

They can act by:

(1) contact killing only the parts of the plant touched;

(2) leaf translocation to kill whole plant—most herbicides used in forestry are such systemic herbicides.

Persistent herbicides, acting through the soil, are applied after planting to keep the soil clear either by preventing emergence of herbaceous weeds or as post-emergence herbicides.

After planting, selective treatments are necessary to avoid damaging young trees. Selectivity is relative, not absolute, and means that under a given set of conditions certain species are killed or seriously injured whereas other plants are not injured. However, a given herbicide is selective only within certain limits of dose, environmental condition, method, and season of application. It is possible to injure or kill the crop if the selective herbicide is not used properly. They include:

1. Systemic, foliar, and non-persistent herbicides. These are the most efficient at controlling existing perennial weeds.

2. Persistent herbicides that are sprayed on clean soil to kill herbaceous vegetation. Their efficiency and selectivity are very dependent on soil conditions and rainfall and on an even application.

3. Persistent foliar herbicides. These are very efficient for controlling weeds after emergence.

Herbicides are applied by:

(1) ultra-low-volume sprayers, especially controlled droplet applicators which require very little water; they can be used only where spray does not drift much;

(2) medium- or high-volume knapsack sprayers;

(3) granule applicators;

(4) tree injection;

(5) surface wiping of foliage using a wick-like applicator.

The timing of application and area treated around each tree are important attributes of success. For controlling weeds round individual trees, a spot application of 1 m diameter or a strip along planting rows 1 m wide is normally perfectly adequate. Spot applications usually treat only 25 per cent of the total area.

In spite of the fact that: (1) correctly applied herbicides have been proved to be the cheapest and most efficient way of controlling weeds (Barthod *et al.* 1990), (2) compared with the total amount of herbicides used in agriculture, the amount of herbicides used by the forest industry is insignificant (0.06 per cent in the United Kingdom in 1988 (Williamson 1990), and (3) the forest area treated by herbicides is extremely low (0.3 per cent of the French forest area is treated annually (Frochot 1992)), the use of herbicides in forestry is likely to decrease in most European countries with increased environmental pressures. There are important differences between European countries concerning the authorized use of herbicides: northern European regulations are very restrictive (in Sweden the application of herbicides is generally forbidden), southern European countries and the United Kingdom regulations are less restrictive, but there is a trend towards a common regulation inside the European Union.

The future of forest vegetation management is tending towards integrated management, including herbicide treatments in a more global strategy of environmentally and economically sound silvicultural practices, which include:

◆ soil preparation;

◆ manual weed control;

◆ mechanical treatments;

◆ individual tree protection;

◆ possibly controlled grazing and prescribed burning.

9 Nutrition and fertilizers

Attitudes towards the use of fertilizers in forests have changed considerably during the twentieth century, largely as a result of the great increase in planting and intensification of forest management. In the late nineteenth and early twentieth centuries the accepted view that 'almost any soil can furnish a sufficient quantity of mineral substances for the production of trees' (Schlich 1899) contrasts with the now commonplace use of fertilizers to correct deficiencies. Miller (1981*a*) contends that there is no great disparity between these outlooks. An understanding of nutrient cycling in forests has shown that while fertilizers can do much for forests before canopy closure, thereafter the nineteenth century ideas, which related to predominantly closed canopy forests, remain true. Such forests have efficient nutrient-conserving mechanisms but during the years before the canopy closes, while trees are accumulating nutrients to build a canopy, growth is dependent on the availability of soil nutrients. It is during this period that fertilizing and other practices developed in agriculture can be used by foresters to improve tree growth.

Nutrient requirements of trees

Plant growth can be substantially reduced if there is an insufficient supply of any of at least 12 nutrient elements listed in Table 9.1. These exclude carbon, hydrogen, and oxygen, which are derived from the air and from water. Others, such as sodium and silicon, may be beneficial under some circumstances and for some species. The importance of each is discussed by Mengel and Kirkby (1978) among others. When a nutrient, or indeed any other factor such as water or light, is present at a critical level it can assume enormous importance and may far outweigh all other factors in determining the health and growth rate of a crop. A recognition of this led Mitscherlich (1921) to propound his 'Law of minimum' which, as stated by Baker (1934), is that 'bringing any factor nearer to the optimum level will increase the yield of a crop, but increasing factors that are markedly deficient will increase the yield disproportionately.' It is helpful in silviculture, and especially in the field of nutrition, to look for such limiting factors when confronted with the many that are integrated to produce a certain growth or yield since by modifying these the greatest improvements can be made.

Table 9.1 Ranges of nutrient concentrations found in the leaves of pines and spruces during the dormant season (modified from Taylor 1991 and Watts 1983)

Element	Macronutrients Concentration of nutrient element (% dry weight of foliage)		
	Severe deficiency	Slight/moderate deficiency	Adequate
N	1.0	1.1–1.2	1.5
P	0.08	0.14–1.5	0.18
K	0.3	0.3–0.5	0.5–0.7
Ca	0.05	0.07	0.1–0.2
Mg	0.03	0.06	0.12
S	0.05	0.14	0.16
	Micronutrients Concentration of nutrient element (p.p.m. of dry foliage)		
	Deficiency likely		Adequate
Mn	4		25
Fe	20		50
Zn	9		15
Cu	2.5		4
B	5		20
Mo	0.1		0.3

Recent investigations of ecosystem processes, especially those that lead to the transfer, transformation, or accumulation of nutrient elements, have contributed greatly to knowledge of forest nutrition (e.g. Cole and Rapp 1981; Duvigneaud 1985; Ranger *et al.* 1991; Hüttl *et al.* 1995; Likens and Bormann 1995). Few have contributed more to this than Miller (1979, 1981*a,b,c*, 1995) and his work is drawn upon here. An essential characteristic of forests is the development of a distinct forest floor that results from the return, through litterfall, of leaves and other tree debris. Two-thirds to three-quarters of the nutrients absorbed from the soil and atmosphere each year are returned to the soil in the litter. They eventually become available for re-use by the trees. A detailed account of the various cycles within the ecosystem, whereby nutrients move between the living vegetation, the soil organic layers, and the lower soil horizons, is given by Miller (1979).

Nutrient cycles occur at three interrelated levels:

1. The geochemical cycle in which nutrients enter a forest site in rainwater, as aerosols, as dust trapped on tree surfaces, through absorption of gases, and through biological fixation of nitrogen.

2. The cycle transferring nutrients from zones of accumulation within the ecosystem.

3. The physiological cycle within the tree, moving recent uptake into temporary storage and subsequently mobilizing it during the growing season or withdrawing it from ageing tissues.

Mycorrhizas

The important functions in tree nutrition of the fungal associations with roots, known as mycorrhizas, must be recognized. They have been described and discussed in detail by Marx (1977), Harley and Smith (1983), Bowen (1984), and Garbaye (1991). Mycorrhizas on trees are of two main types—ectomycorrhizas and vesicular-arbuscular mycorrhizas. Most plantation species form ectomycorrhizas that infect and envelop the short lateral roots. The hyphae from these grow into the soil and the litter. Vesicular-arbuscular mycorrhizas form on eucalypts, and a few other tree genera, and differ from the ectomycorrhizas in that they do not form sheaths round the roots though they too penetrate the soil and litter. Mycorrhizas act as huge but very fine extensions to the root systems and facilitate the uptake of many nutrients, particularly poorly mobile ones such as phosphorus, and sometimes water. Sanders and Tinker (1973) have shown that the inflow of phosphorus into mycorrhizal plants can be four to five times that into non-mycorrhizal plants. Mycorrhizas are particularly important in soils where nutrients are in low supply, or in relatively inaccessible forms. They might also be valuable as an alternative means of nutrient uptake in a range of soils that are unfavourable to root growth—including very acid soils, those that experience extremes of temperature, high aluminium, and high salinity (Bowen 1984).

Both types of mycorrhizas exist commonly in most soils and indigenous trees usually become infected with an appropriate fungus, or fungi, naturally. Where suitable fungi are absent, spectacular increases in growth have occurred following their introduction, for example in pines that have been introduced to parts of the world where they are not endemic. Where indigenous fungi are only poorly effective, inoculation can lead to dramatically better growth, though the problem is to select an effective fungus that will displace or compete favourably with the indigenous one. This is easy to do in nurseries where the soil is commonly sterilized (Le Tacon *et al.* 1988; Garbaye 1990; Le Tacon and Bouchard 1991), but the conditions which influence inoculation and persistence in the field are far less well known. Much research still remains to be done before mycorrhizas can be managed effectively on a large scale.

Nutrition of broadleaved and coniferous species

Miller (1984) has shown that deciduous broadleaved species do not have very different nutrient demands from evergreen conifers of the same growth rate, yet broadleaved trees are widely accepted as being nutrient demanding. He contends that they are site- rather than nutrient-demanding, being relatively unable to obtain nutrients from intractable soil sources, rather than requiring larger amounts of nutrients than conifers. Broadleaves are therefore more commonly associated with 'fertile' sites where nutrients are present in relatively easily available forms.

The rate of nutrient cycling differs markedly between deciduous and evergreen species. Cole (1986) has demonstrated that in northern latitudes decomposition is slow, especially in coniferous forests, and this results in the accumulation of large amounts of organic matter and nutrients in the forest floor layer. Seventy per cent of the above-ground organic matter in boreal forests is on the forest floor; this drops to about 15 per cent in temperate forests, and to only 3 per cent in tropical rain forests. The distribution of nutrients between the living vegetation and the dead organic matter on the forest floor follows approximately the same pattern.

The nutrient status of the litter from deciduous trees is much greater than coniferous litter, and this leads typically to far more rapid decomposition rates. Because the foliage of deciduous species is replaced each year, unlike that of evergreens, the cycling of nutrients also differs between the two groups. Litterfall is much greater in deciduous forests, and the annual uptake, requirement and return of nutrients is far greater for most elements in deciduous than in evergreen ecosystems, as shown in Table 9.2. Only in the case of phosphorus is the difference relatively small. Table 9.2 shows that deciduous species translocate more nitrogen than evergreens: about a third of the annual requirement of deciduous trees is met through translocation from old to new tissues.

Forest fertilizing—principles

As a result of investigations into ecosystem processes, Miller (1981*b*) has proposed three ideas relating to forest fertilization that have done much to rationalize the huge amount of sometimes apparently contradictory information on the potential role of fertilizers in forests.

Table 9.2 Comparison of deciduous and evergreen species in relation to the uptake, requirement, and return of five macronutrients (kg ha^{-1} yr^{-1}) (from Cole 1986)

Element	Uptake		Requirement		Return	
	Deciduous	Evergreen	Deciduous	Evergreen	Deciduous	Evergreen
Nitrogen	70	39	94	39	57	30
Potassium	48	25	46	22	40	20
Calcium	84	35	54	16	67	29
Magnesium	13	6	10	4	11	4
Phosphorus	6	5	7	4	4	4

(1) Generally fertilizers benefit the trees and other vegetation, not the site

Following application, fertilizer nutrients rapidly become distributed between the trees, the ground vegetation, the forest floor, and mineral soil horizons. Small amounts may also be lost through leaching, or in the case of nitrogen, gaseous diffusion, but the forest floor and surface horizons are remarkably effective in retaining received nutrients, more so than under other forms of vegetation. If the trees are suffering from a deficiency of an applied nutrient, they will show a response, typically for 5 to 10 years (e.g. Pettersson 1994). In deciding the nature and timing of fertilizer application it is important to know whether the additional amount of nutrient that promotes the improved growth is derived from nutrient reserves stored in the tree or in the soil or if it is the result of a more rapid cycle between the two.

There are normally two response patterns. The first (Figs 9.1 and 9.2) is shown by all fertilizer treatments and consists of a rapid increase in growth, followed by a slow and progressive decline to near preapplication growth rates. In the years immediately after fertilizer application, the trees make no greater demands on the soil and organic matter than in unfertilized areas unless rooting is greatly enhanced. Improved growth is therefore normally explicable in terms of nutrients accumulated in the trees' tissues immediately after fertilizer application in excess of requirements for growth. This explains why improved growth continues in the years after treatment, at least until the time when concentrations in the trees' tissues fall to pretreatment levels.

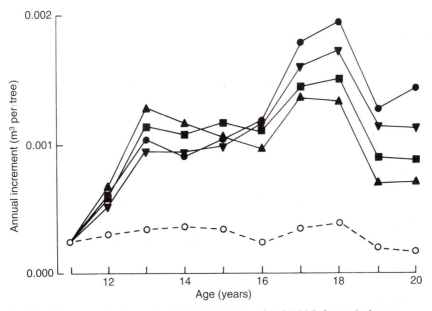

Fig. 9.1 Volume growth of young Corsican pine (*Pinus nigra* subsp. *laricio*) (before peak of current annual increment) showing that once response to fertilizer treatments (solid lines) has finished, growth rate remains above the level of unfertilized trees (dashed line). By age 20, nitrogen concentrations in the foliage were no longer different between treatments. (From Miller 1981*b*.)

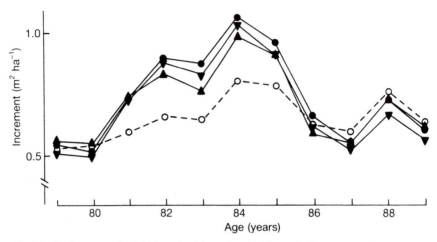

Fig. 9.2 Basal area growth of old Scots pine (*Pinus sylvestris*) (after peak of current annual increment) illustrating that once response to fertilizer treatments (solid lines) has finished, growth rate falls below the level of that in unfertilized trees (dashed line). (From Miller 1981*b*.)

Thus, it is the trees, rather than the site itself, that normally benefit from fertilizers. Applications should be timed to maximize both the rate of uptake and the storage within the trees. Because much of the storage occurs in foliage, it is important not to allow crowns to become too sparse before applying fertilizers.

Only on some heavily fertilized sites does the second pattern sometimes emerge in which growth remains consistently faster than in less heavily treated areas. In such cases, the amount of applied nutrient retained in the site is large in relation to its original capital. This results in a long-term response.

(2) Response to fertilizers is best considered as a reduction in rotation length

There is little to contradict the idea that fertilizer response can be adequately described using the simple analogy of an acceleration through time. This is because crops eventually rejoin the growth curve (Fig. 9.3) at a point commensurate with their development stage, or size, rather than chronological age. They jump several years ahead and eventually grow faster, or slower, than a similar untreated crop depending upon which side of the growth curve is being considered. Depressions in growth of fertilized trees are commonly observed in older crops once the response period has ended (Fig. 9.2). Similarly in young crops, growth never falls to the level of that of unfertilized trees on an age basis (Fig. 9.1), but this does not mean that there is a continuing long-term response. Apart from possible effects on wood properties, for example a reduction in specific gravity (Heilman *et al.* 1982*a*), there is no evidence that a fertilized crop will differ significantly from an unfertilized one that is grown over a longer rotation.

(3) Fertilizer requirements alter with development stage

There are three distinct stages in the life cycle of a plantation crop (Fig. 9.4). In Stage I the crowns are developing and need large amounts of all nutrients, few of

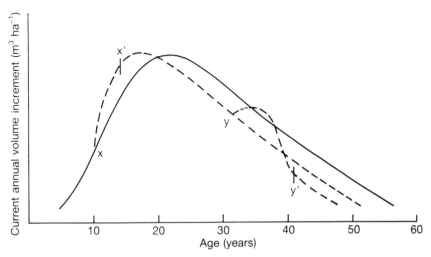

Fig. 9.3 Hypothetical growth curves illustrating that trees fertilized at points x and y will, once response to fertilizer has finished, follow growth patterns (dashed lines) parallel to, but ahead of, the curve for the unfertilized trees (solid line) from points x′ and y′.

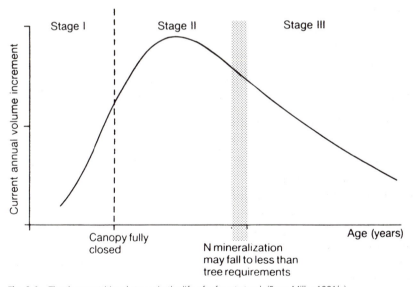

Fig. 9.4 The three nutritional stages in the life of a forest stand. (From Miller 1981*b*.)

which are returned through litterfall. During most of this period the trees are not fully using the site. Much of the soil volume and a large part of the nutrient input in rainwater is taken up by weeds (see Chapter 8). Although, in most soils, the amounts of potentially available nutrients are relatively large compared with the amount in the crop, and although most trees are able to obtain nutrients from fairly intractable soil sources, there are many soils in which one or more nutrients may be in very low supply. This is, therefore, the stage when the supply of available soil nutrients is likely to be the critical factor controlling the rate of crown development, and response to a range of fertilizer nutrients may then be expected. It should be noted that if weed control is inadequate, most of the applied nutrients will be taken up by the weeds which will then compete even more strongly with the trees.

By Stage II, which begins at the time the canopy closes, immobilization is reduced and continues only in the trees' woody components. Although the uptake of nutrients rises to high values and tends to come to a peak shortly after the age of maximum current annual increment, most are recycled. During this period nutrient cycling, both within the ecosystem as litterfall, crown leaching, root death, and root exudation, and within the trees themselves, becomes the dominant process. It is also the time when the capture and retention by the tree crowns of atmospheric nutrients in rainwater, aerosols, dust, and gases is most efficient. These natural inputs may balance losses from the mean annual removal of wood, at least for elements other than nitrogen and phosphorus (Miller 1979). Improved cycling of nutrients and enhanced input means that for many elements the demands made on the soil nutrient capital may be very low. Demands are often so low that supplies are adequate and hence responses to fertilizers are unlikely.

Anything in Stage II or later that reduces foliage biomass, such as thinning, may cause a partial reversion to Stage I. A crop may therefore respond to fertilizing with nutrients originally found to be deficient in Stage I. So, even in Stage II which is generally unresponsive, fertilizers can be used on deficient sites to accelerate recovery from thinning or perhaps insect defoliation, though if there was no serious deficiency in Stage I no response to additional fertilizer would be expected.

During Stage III nutrient immobilization continues at a low rate within the trees but the increasing rate of immobilization in the soil organic layers may sometimes be much greater than input, particularly beneath some species of conifers and on very acid soils. This is particularly characteristic of northern coniferous forests. For nitrogen, at least, the quantity that is removed from the cycle in this way can exceed the amount accumulated within the trees. This immobilization can lead, on sites of low capital, to nitrogen deficiencies later in the rotation. Eventually, at an advanced age, a crop's demands for nitrogen may fall below the rate of mineral nitrogen production and a natural recovery from a deficiency may be observed.

The time at which a crop enters Stage III depends upon the available nitrogen capital of the site. On sites where there is little it may set in soon after the canopy closes with virtually no intervening Stage II. Where the nitrogen capital is higher, deficiency will appear much later. On many sites it does not appear at all.

A fourth stage has been proposed by Heal et al. (1982) which occurs immediately after clear-felling. At this time, a high input of crop residues, cessation of tree uptake, and initially, little uptake by a sparse ground flora, allows decomposition processes to predominate. Nutrient retention is at a minimum through lack of plant uptake and increased runoff. Nutrient, and especially nitrogen, losses may be increased considerably, especially in areas favoured by high decomposition rates. A prudent forester takes advantage of these available nutrients by planting a successor crop quickly after felling.

Thus, fertilizers are most likely to be useful before the canopy closes or towards the end of a rotation. During the early years when the trees are very dependent on soil supplies, almost any nutrient element may be deficient. Although tree roots are not exploiting the site efficiently and uptake and requirements are at their lowest, this is the time when judicious fertilizer application can most influence the subsequent development of a stand, perhaps reducing the rotation length by many years. Late rotation deficiencies are, by contrast, almost entirely of nitrogen.

In short-rotation forestry, such as coppice, harvesting occurs at Stage II and large amounts of nutrients can be exported from the site (see Chapter 15). Depending upon what is removed—whole trees or only larger stems—the requirement for replacing nutrients can vary enormously (Ranger and Nys 1996).

Forest fertilizing in practice

In natural ecosystems the minerals available for nutrition represent the balance between the nutrient flux density in the soil, the uptake rate by the crop, and immobilization by competitive uptake and chemical binding (Ingestad 1991). Given this balance, it is obvious that fertilizers should, ideally, be introduced into an ecosystem continuously to supplement the mineralization processes. The rate of addition should equal the potential rates of uptake and binding processes. At present it is quite impossible to achieve this sort of precision in the vast areas of forest which need added nutrients. In practice, typical fertilizer doses are very large and applied at widely spaced time intervals in comparison to the continuous natural fluxes. Current practices are described fully by Ballard (1986) and Bonneau (1986).

Though fertilizers have been investigated and used to correct nutrient deficiencies since the mid 1800s in parts of Europe (Baule and Fricke 1970), it was largely in countries concerned with establishing new forests on bare and infertile sites that advances in practice were made. Thus, in Europe, the United Kingdom was among the first to apply nutrients on a large operational scale to correct Miller's (1981*b*) Stage I deficiencies. Sweden and Finland have been among the leaders in correcting Stage III nitrogen deficiencies since the 1960s. Forest fertilizing has been widely practised since the mid-twentieth century. It is normally the major nutrients, nitrogen, phosphorus, and potassium, that are deficient, and of these, phosphorus is normally by far the most significant in improving growth during the early years of a crop's life (Taylor 1991). This is because many rocks and soils are naturally deficient in available phosphorus.

Though nitrogen is one of the most abundant elements in forest ecosystems, it is not contained in any soil minerals and most must be supplied from decaying vegetation. Stage III deficiencies occur commonly in coniferous forests in cool temperate climates where deep organic layers develop as a result of slow rates of litter decomposition. The possibilities of hastening decomposition by applying lime to promote more biological activity have shown promise on an experimental scale (Dickson 1977), but the quantities required, at 5 to 10 t ha^{-1}, make it both difficult and expensive. In France and Germany, liming is sometimes done to improve the biological properties of acid soils so that they are capable of carrying stands of broadleaved trees rather than only conifers (Baule and Fricke 1970). Nitrogen deficiencies also occur on sandy heaths, especially well-ploughed ones where mineralization of organic debris is faster than the supply, and the most extreme problems are found on 'soils' such as sand dunes and some open cast mining sites where there is virtually no organic material present.

Potassium deficiencies become most obvious in vigorous young stands, especially where a nutrient imbalance has been created after applying nitrogen or phosphate. Frequently symptoms disappear naturally as the excess of other nutrients becomes reduced.

Among other nutrients, magnesium deficiencies arise occasionally, especially in nurseries. Copper deficiencies are found on some sites where organic nitrogen mineralizes very fast, such as some cultivated, sandy soils and on some peats (Binns *et al.* 1980). *Pseudotsuga menziesii* is particularly sensitive to copper deficiency. Iron and manganese deficiencies occasionally appear on highly calcareous soils, especially old arable sites, such as in *Pinus* and *Fagus sylvatica* mixtures on chalk downland in southern England. Deficiencies of other elements, such as boron, are known in many temperate countries, and if they are common it is usual to add small quantities to other fertilizers. For example 0.2 per cent boron is added to all fertilizers applied in Sweden. Very often deficiencies of more than one element occur at the same time, either because the site is deficient or because the application of one element may cause an imbalance and induce a deficiency of another. Such a situation arises in south-west France where applications of copper alone can be toxic to *Pinus pinaster*, but when applied with phosphorus good growth is obtained (Saur 1993).

Fertilizers are not normally needed in large amounts, if at all, for second and subsequent rotations because the debris left after harvesting—leaves, branches, tops, and the litter layer—contain most of the nutrients from the previous crop. Little is lost in normal harvesting systems provided leaching is not excessive before the new crop becomes established, though whole-tree harvesting can remove considerable quantities of nutrients and supplements are then needed.

Diagnosis of nutrient deficiencies

Visual symptoms

Serious deficiencies of many nutrients can be recognized by visual symptoms of leaf colour or other features, particularly during the dormant season with evergreen conifers. The most common symptoms are summarized in Table 9.3 and have been described and illustrated by Taylor (1991) among others. The recognition of deficiencies in deciduous trees is more difficult and less well documented, partly because they tend to be grown on sites where nutrition is not a problem. Baule and Fricke (1970) illustrate some deficiencies of broadleaved trees well.

Growth responses to an added nutrient can normally be obtained long before a visual symptom of deficiency is apparent, so prudent managers normally attempt to anticipate such serious conditions by other methods.

Chemical analysis of plant tissues

The concentration of elements within plant tissues can often reflect the nutrient status and health of a crop. Developing deficiencies can be detected by chemical analyses of tree tissues, using techniques developed by Leyton (1958) and Bonneau (1978). Conventionally, with evergreen conifers, samples of the youngest foliage from an upper whorl of branches are analysed. They are usually collected during winter, when foliar nutrient concentrations are fairly stable. There is less

Table 9.3 Visual symptoms of some nutrient deficiencies in evergreen conifers

Element	Visual symptoms
N	Usual dark-green colour of healthy foliage becomes *uniformly* paler, often yellow-green or yellow over the whole crown. Needles shorter and lighter in weight than healthy foliage. Height growth *suddenly* reduced. Leading shoot spindly. Form of tree unchanged.
P	No marked discoloration but needles dark, dull green. (P- and N-deficiencies often occur together in which case P-deficiency symptoms may be masked by N-deficiency symptoms.) Needles are much reduced in weight and length; height growth reduces slowly and form of tree is unchanged except after prolonged deficiency when apical dominance may be lost.
K	With moderate deficiency in young trees, tips of needles at *top* of current shoots (or previous year's shoots in some pines) become yellow. In more severe cases, all needles on current year's shoots affected. On older trees, yellow tips are confined to older needles, often on lower branches only. Size of needles similar to that in healthy foliage. Growth not always significantly reduced except in cases of extreme deficiency and in very young trees. Form of trees unaffected, except in severe cases when apical dominance may be lost and a stunted, bushy habit develops, especially in very young trees.
Mg	Current year's needles become yellow from *base* of shoot upwards and yellowing spreads from needle tip downwards except in pines where all current needles yellow together. Often confused with K deficiency but transition from yellow to green is much sharper with Mg deficiency.

Sources: Binns *et al.* (1980); van Goor (1970).

experience with deciduous species because they are more commonly planted on relatively nutrient-rich sites, but sampling is usually done in July or August, the middle of the growing season.

The concentrations required for optimum growth vary between parts of a tree and between species and even provenances. Optimum concentrations also differ according to which parameter of growth is being considered. For example the optimum for growth in height is often at a lower concentration than for growth in volume. Concentrations also vary significantly with the size or age of the plant (Miller *et al.* 1981; Miller 1984). However, concentrations tend to be relatively uniform for a particular development stage and type of tissue, such as leaves within a species or even a group of species.

A deficiency may be considered to exist if the addition of an element to the soil–plant system results in an increase in growth (Prichett 1979). However, it is not easy to define what critical, or especially optimum, concentrations are because levels that are adequate for tree growth in one set of nutritional and environmental conditions may be inadequate in others. Thus, a critical concentration is an ill-defined value and may more accurately be considered as a narrow range of concentrations. Optimum concentrations tend to have a much wider range. An indication of the common ranges of concentrations is given in Table 9.1 (p. 122). It is also important that the concentrations of some nutrients should be in reasonable balance with those of others. If, for example, there is a substantial departure from the ratios of N:P:K of 10:1:5, this indicates possible nutritional problems.

Bioassays

Methods developed by Jones *et al.* (1991) and McDonald *et al.* (1991) are showing potential for providing rapid tests for both nitrogen and phosphorus status of trees, using newly excised roots. They involve measuring the pattern or rate of uptake of ^{15}N or ^{32}P from solutions in which the roots are placed.

Soil analysis

Except in nurseries, equivalent techniques to those used in agriculture for detecting likely deficiencies by analysing samples of soil have proved difficult in forestry (Leyton 1958; Khanna 1981). This is partly because it is not possible to distinguish between total and available nutrient concentrations with current techniques, especially in soils with large amounts of organic matter that often predominate in forests, and in which deficiencies are common. Further complications arise because of soil heterogeneity and uncertainties about which soil horizons to sample because trees often root deeply and extensively throughout the profile (Meredieu *et al.* 1996). There is also a lack of information for interpreting test results in terms of tree responses to fertilizers (Prichett 1979). The level of nitrogen in the humus has been suggested as a good measure of nitrogen status by Miller *et al.* (1977).

Fertilizer regimes

Experimental and other evidence of treating deficient sites has led to the development of recommended fertilizer regimes for different species and sites. The regime selected depends upon the finance available, among other factors. Normally it is the one at which the investment return is maximized. Some examples are given by Taylor (1991).

Expensive, high-input regimes necessitate relatively high and valuable outputs to justify the costs. There are frequent debates as to whether less expensive regimes based on lower inputs or less demanding and less valuable species might be more profitable especially where high risks to the crop exist from windthrow, fire, or grazing. An additional problem arises if complete trees are harvested from a site. Since two-thirds to three-quarters of the nutrients in a tree are contained in its leaves (Miller 1995), losses can be considerable. Soil weathering processes and atmospheric inputs are often insufficient to compensate for the removals, so that fertilizers must be added to maintain levels of production especially if rotations are very short, as with poplar coppice (Ranger and Nys 1996).

Effects of fertilizers on wood quality

The usual purpose of adding fertilizers to plantations is to improve their economic performance by increasing the rate of volume production. The immediate response to increasing the level of a deficient nutrient is usually an increase in the photosynthetic area that then causes the tree to grow faster. The faster growth can affect some wood properties in exactly the same way as, for example, thinning or an initial wide planting spacing might. These aspects are discussed in Chapter 10, and have been reviewed in relation to fertilizers by Bevege (1984). The effects of increased growth rates on wood properties, especially in conifers, are complex and generalizations can be unwise. To the extent that they are possible at all, there is evidence for a reduction in the proportion of latewood and consequently in wood density; a tendency for fibre lengths to be decreased, though the evidence for this is somewhat contradictory. If Stage I fertilizing is associated with relatively wide spacings and short rotations, the proportion of juvenile wood in the stem is increased. All these effects reduce the quality of the wood for some uses, though increased yields normally more than compensate for such reductions, in terms of the total value of the crop. Only when the application of a deficient nutrient improves the form of a tree by, for example preventing stem dieback, does fertilizing unequivocally improve both yield and quality.

Types of fertilizers

Ideally, the fertilizers used in forests should be cheap per unit of nutrient, concentrated to allow low application costs from ground machines or the air, long-lasting, and unlikely to cause any pollution (Binns 1975). Those used widely in forestry are shown in Table 9.4. Detailed accounts of these fertilizers and their properties are given by Taylor (1991) and Prichett (1979).

Nitrogen (N)

In current world-wide practice only three water-soluble fertilizers, ammonium nitrate, calcium ammonium nitrate, and urea, are used to any significant extent. All are concentrated and cheap in terms of the amount of nitrogen applied. There is evidence that in northern latitudes, including Scandinavia and Britain, ammonium nitrate is the better source for getting nitrogen into the trees (Binns 1975; Taylor 1991). In Britain, the differences in responses between this and urea in terms of equivalent amounts of nitrogen are small so that in practice urea is usually favoured because it is cheaper and more concentrated, which makes it more economical to apply. By contrast, in Sweden ammonium nitrate proved more effective and was applied in the 1970s, though in the 1980s a change was made to calcium ammonium nitrate in an attempt to counteract soil acidification caused by acid rain. Slow-release sources of nitrogen such as urea formaldehyde and calcium cyanamide, do not usually compare in effectiveness with water-soluble sources and they are generally too expensive to use in any case. Nitrogen is usually applied to crops at rates of about 150 to 250 kg of element N ha^{-1}. If nitrogen is applied to young plants, care must be taken not to let the concentrated fertilizer touch the roots as this can kill the trees.

There are, of course, also possibilities of using nitrogen-fixing plants to improve nitrogen nutrition, and use is being made of this technique in New Zealand, among other places (Miller 1981c). While they can sometimes be useful for correcting a

Table 9.4 Commonly used nitrogen, phosphorus, and potassium fertilizers

Fertilizer		Nutrient element (%)
Nitrogen		
Urea	$(NH_2)_2CO$	46
Ammonium nitrate	NH_4NO_3	34.5
Calcium ammonium nitrate	$NH_4NO_3 + CaCO_3$	27.4
Ammonium sulphate	$(NH_4)_2SO_4$	20.6
Phosphorus		
Rock phosphate	usually $Ca_{10}F_2(PO_4)_6 \times CaCO_3$	11–17
Superphosphate	$CaH_4(PO_4)_2.H_2O + CaSO_4$	8–9
Triplesuperphosphate	$CaH_4(PO_4)_2.H_2O$	19–21
Potassium		
Potassium chloride	KCl	50
Potassium sulphate	K_2SO_4	42

Stage I deficiency, such as on sand dunes, they can not easily be used for Stage III because nitrogen-fixing plants are too light demanding to survive under forest canopies. Most will not grow in the acid soil conditions which generally prevail where deep organic layers develop. The few species that might grow on such soils, such as broom (*Cytisus scoparius*) are difficult to establish and to protect from browsing, though if they do become established they often grow so vigorously that they compete strongly with the trees.

Phosphorus (P)

Phosphorus is usually applied at rates of 50 to 125 kg of element P ha^{-1} at planting. The choice is principally between water soluble materials, such as superphosphate and triplesuperphosphate, and slowly soluble rock phosphate.

Rock phosphate, normally of north African origin, has proved to be of particular value on acid soils with a pH below 5.5 (Binns 1975) and is widely used. On soils with high phosphorus-fixing capacities, such as those derived from calcareous or serpentine rocks, rock phosphate becomes less available and so the more soluble forms such as superphosphate are appropriate.

Potassium (K)

Of the two common forms of potassium (KCl and K_2SO_4), only potassium chloride is used to any extent because it is cheap and easily available. It is usually applied at rates of about 125 kg of element K ha^{-1}. Potassium sulphate is of value when the absence of further chlorine should be avoided or the addition of sulphur is important.

Mixed fertilizers

Extensive forest areas quite often need to be treated with both phosphorus and potassium and more rarely with phosphorus and nitrogen. Where phosphorus and potassium are needed together, physical mixtures of rock phosphate and potassium chloride are difficult to spread uniformly. Granular fertilizers consisting of both materials have better spreading qualities. Where urea and rock phosphate are required on the same area, it is usually thought best to apply them in two separate operations since their spreading characteristics are quite different.

Sewage sludge and animal slurry

Nutrients contained in sewage sludge and animal slurry can benefit tree crops in the same way as more concentrated natural or manufactured fertilizers. There are several examples of successful trials in the literature (e.g. McAllister and Savill 1977; Taylor 1991). The enormous bulk of the material makes the costs of transport and application very high (Fig. 9.5). It would require, for example, over 100 t

Fig. 9.5 Applying sewage sludge. (Crown Copyright.)

of slurry to supply the amount of phosphorus contained in half a tonne of rock phosphate which is a quantity commonly used per hectare on deficient sites. With the same weight of rock phosphate 200 ha could be treated. Guidelines for the application of sewage sludge to forests have been provided by Wolstenholme *et al.* (1992). They include a maximum recommended volume of 200 m^3 ha^{-1} $year^{-1}$, and less on shallow soils; the need for screening or macerating the sludge, avoiding application on slopes greater than 25° and reducing applications on less steep slopes, and not exceeding specified application limits where various potentially toxic heavy metals are present.

Methods of fertilizer application

Before the 1960s it was common to apply fertilizers by hand, as spot treatments in and/or around the planting holes. Subsequently it was found that broadcast applications over the whole of the site to be exploited by tree roots resulted in significantly better growth (Dickson 1971). Since then techniques, first for ground application by machines and later for fixed wing aircraft and helicopter application (Fig. 9.6), improved and became much cheaper. By 1970, most fertilizer was spread by one of these two methods rather than by hand, and since the 1990s helicopters have predominated. Progress with aerial methods has been opportune because many young plantations are virtually impenetrable from the ground. There are still problems in

Fig. 9.6 Applying rock phosphate from the air. (Crown Copyright.)

obtaining acceptably uniform rates of spread from the air, though improvements in the form of some fertilizers used and the techniques of application have been considerable. A wide and even spread has been achieved through the use of large (5–9 mm diameter) granules in Sweden. If a relatively even spread can be assured, application rates can be reduced.

Season of application

At one time it was believed that highly soluble fertilizers, including all commonly used sources of nitrogen and potassium, could easily be lost from a site by leaching if they were not applied during the active growing season, usually March to September. Recent research (e.g. Heilman *et al.* 1982*b*; Taylor 1991) has indicated that season of application has little influence on the effectiveness of any of the commonly used fertilizers, provided they are not applied to frozen or snow-covered ground.

Environmental implications

Indiscriminate application of fertilizers can result in the enrichment of drainage water and consequent eutrophication of streams, lakes, and reservoirs. In some circumstances algal blooms may develop. These can block filtering equipment from reservoirs and the reduced oxygen levels associated with the algal growth can cause the death of fish and other aquatic life. Fertilizing non-forest land can also enrich and destroy various habitats of rare plants, for example on boglands. Clearly guidelines for preventing such environmental damage should be followed and some have been given by Binns (1975). They include measures to minimize the contamination of water and sites of value and the avoidance of excessive soil enrichment by accurate and uniform application. In some countries, such as Sweden, fears of soil acidification by some nitrogen fertilizers have lead to the recommendation that only calcium ammonium nitrate is used.

10 Spacing, thinning, pruning, and rotation length

The spacing between trees or stocking at which a crop grows affects the degree of competition in the stand. This influences mortality, total production per unit area, the size to which individual trees will grow, several aspects of wood quality, and sometimes susceptibility to pests and diseases. The value of timber and costs of management practices are both affected by spacing. A clear understanding of the effects of possible spacing and thinning regimes on production is one of the most important aspects of silviculture since they influence the end use of the timber and its profitability, and, as with species selection, are directly under a forester's control.

Stand density relationships

Maximum stand density

In pure even-aged stands a site can support only a certain number of trees of a given size. Once it is fully occupied self-thinning begins to occur as individuals compete for resources. The maximum number which can be carried depends upon species and stage of development. Stand densities beyond a certain level can never be realized, however close the initial spacing may have been. As a stand becomes older and crown size increases, the upper limits of stand density decrease.

At high stand densities that result in tree mortality, the relationship between average tree size and number per unit area can be represented graphically by a single line, irrespective of differences in age, site quality, or initial spacing. The reasons trees die in conditions of severe competition lie in genetically controlled, physiological processes—for example, the minimum photosynthetic area required to sustain a stem of a given. Though the environment can, and clearly does, affect growth rate, it does not alter maximum average plant size for a given stand density, which is a phenomenon independent of age (Drew and Flewelling 1977).

Thus, Reineke (1933) found that the relationship between maximum number of trees and average diameter can be represented by a straight line when plotted on logarithmic scales, and that for most species the slope of this line is constant, though its elevation (k) differs very slightly between species. Reineke represented the line by the equation:

$$\log p = -1.605 \log D + k \tag{10.1}$$

where p is the number of trees per unit area, D their average diameter at breast height, and k is a constant that varies with species.

A very similar relationship has been found to be applicable much more widely than to trees alone by Yoda *et al.* (1963). They found that in all the plant species examined, including some herbs, the slope of the log weight–log density line was nearly equal to $-3/2$ irrespective of the differences between species, that is

$$\log w = k \log p^{-3/2} \tag{10.2}$$

where w is the average weight of surviving plants and p is the density (Fig. 10.1). Unlike Reineke, they used stand density as the independent rather than dependent variable. However, this equation is very similar to Reineke's if tree weight rather than diameter is considered. Tree weight is approximately proportional to the 2.5th power of diameter at breast height, so that Reineke's equation can be transformed to:

$$\log w = k \log p^{-1.558} \tag{10.3}$$

which is very close to the $-3/2$ found for other species.

The relationship given in equation 10.2 is found so universally among plants that is has been termed the *$-3/2$ power law of self-thinning*. It is applicable to pure,

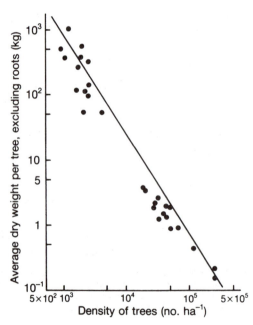

Fig. 10.1 $-3/2$ Power law in natural pure stands of *Abies sachalinensis*. (From Yoda *et al.* 1963.)

even-aged stands in which competition is severe enough for self-thinning to occur, and is discussed in detail by Harper (1982).

A stand at high density where self-thinning is occurring has many relatively small-diameter trees per unit area compared to one at a low density, where the trees are fewer and larger. In forestry, as well as in agriculture and horticulture, it is usually undesirable to maintain crops at spacings which lead to mortality from competition. Thinning is usually practised with the result that if weight is plotted against stand density, a point well to the left of the self-thinning line in Fig. 10.1 is produced. It is never possible for a plant population to be represented to the right of the line. In thinned plantations the steep slope of the –3/2 power line may never be reached, even though competition may be occurring. In spite of competition, as a stand grows and fully occupies the site, the yield of dry matter per unit area rapidly becomes independent of the number of plants present; yield reaches a plateau over a wide range of initial planting distances, as illustrated in Fig. 10.2.

Crown diameter/bole diameter ratio

In considering the growth of individual trees rather than stands, the more space a tree has, the more it will grow in volume (but not in height, as discussed below). Most trees maintain an almost constant ratio of crown diameter to bole diameter throughout the silviculturally critical part of their rotation. The ratio can slightly diminish with age but has never been found to increase (Dawkins 1963). Thus, if

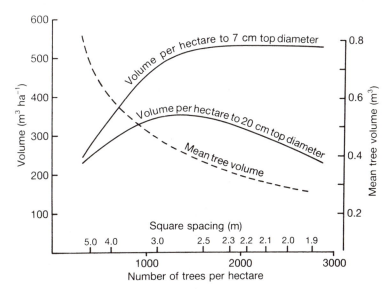

Fig. 10.2 Relationship between spacing and volume production and size distribution in unthinned Sitka spruce (*Picea sitchensis*) of 18.5 m top height. Total production (conventionally production to 7 cm top diameter) becomes independent of number of trees per hectare once densities of about 1500 stems ha⁻¹ have been exceeded. (From Kilpatrick *et al.* 1981.)

the ratio is 20, a healthy tree of mean stem diameter d must, on average, have a crown $20 \times d$ across, and of course, the same mean spacing ($20 \times d$) between trees if it is not to become constricted. A knowledge of this ratio for a particular species enables 'thinning regimes' to be drawn up quite easily, which depend on the diameter to be achieved when serious crown competition begins rather, than age.

Tree height/breast height diameter ratio

The slenderness of a tree stem can conveniently be expressed as the ratio of total height to diameter at breast height (h/d). This ratio is relatively easily manipulated by spacing and thinning. The principle underlying the natural design of most trees is the maintenance of elastic similarity so when the curvature of a stem caused by windsway, or of a branch by its weight, exceeds a specific threshold value, the stem or branch begins to increase in diameter. The mechanism which initiates this response is not known (McMahon 1975).

Trees in very crowded stands have much support from, and contact with, neighbours. They are thin and spindly and have high h/d ratios, usually of more than 100. Trees in more open-grown stands, by contrast, have much greater diameters for any given height, and hence lower h/d ratios. Similarly, slenderness decreases naturally among dominant trees and also in thinned stands but not in unthinned stands, once competition has started. Any sudden exposure of trees in dense stands, for example by thinning, renders them far more susceptible to damage from windthrow and windsnap because they temporarily have an inappropriate h/d ratio, though they adjust by growing in diameter to suit the new spacing over a few years.

Height/density relationships and site index

Unlike plant weight or stem diameter, the average height attained by the dominant trees in a stand is largely independent of density. Height, or conventionally 'top' height, which is the average height of the 100 largest diameter trees per hectare, has been found scarcely to vary at all with density on a given site in numerous species (Evert 1971). It has also been found that within a wide range of conditions, the height to which trees will grow in a given time is more closely related to the capacity of a site to produce wood than any other single measure of trees' dimensions (see Eichhorn's hypothesis, p. 144). Dominant height is consequently used as an index of site quality in even-aged stands of varying densities and silvicultural treatments.

The top height of a stand at a specified age, often 50 years in Europe, is commonly called the site index, and the usual way of determining it requires the use of height/age growth curves to estimate the height at the standard age. Normally, the total range of top heights at this age is divided into a number of classes, each having an equal but limited range of heights (usually 2–3 m). The methods of construction of such curves, and the yield models that are often derived from them, are described by Pardé and Bouchon (1988) and Alder (1980). Top height is not, of course, perfectly correlated with productivity, particularly at the extremes of density. Among broadleaves, such as *Quercus*, that develop rounded tops when mature, decreases in height are known to occur with markedly increased spacing (Savill and Spilsbury 1991). Differences also occur between regions so that, for example, given levels of production are found at lower top heights in windier, western parts of Europe than in more sheltered regions. However, within the normal range of densities at which plantations are established, differences are likely to be small even on quite large regional scales. For example Hägglund (1974) found that height development in *Pinus sylvestris* was comparable in trees from northern and southern Sweden and from thinned and unthinned stands.

Spacing

From the relationships discussed above and in later sections the general effects of spacing on even-aged crops can be summarized as follows:

1. High densities result in the earlier onset of competition between trees and lead to higher mortality through self-thinning.

2. High-density stands generally produce more wood than lower density stands at any age since the resources of the site are fully used earlier in the life of the crop.

3. Once the canopy has closed in widely spaced stands and the site is being fully used, dry matter increment is similar to that in close spacings on the same site; it may even be slightly greater because less is lost through respiration in widely spaced stands.

4. Low densities lead to greater volume production by individual trees because they are subject to less competition.

5. The height growth of dominant trees is little influenced by density.

6. Form factors increase and taper rates decrease in denser crops.

7. Wide spacings lead to larger branches, more knotty timber, a bigger juvenile core, and, in some species, to wood of lower average density.

Pattern of spacing

In most planted forests the distance between rows of trees is usually determined accurately, but on many sites it can be difficult to achieve precision within rows. Rocks, old stumps, and drains often have to be avoided by altering the within-row spacings. The main exception is often seen with species such as *Populus* growing on level, stone-free ground where straight lines can be seen in a number of directions.

For practical reasons, such as ease of finding young trees when weeding and for later line thinning it is important that rows are straight. The rectangularity of spacings, determined by the distance between rows of trees relative to the within-row spacing also has important practical implications. For example, weed control can sometimes be carried out more cheaply by tractor-mounted equipment, so it is sometimes desirable to have enough space between rows to accommodate tractors (at least 2.4 m). To achieve a desired number per hectare, it may therefore be necessary to plant more closely within rows than between them, giving a rectangular spacing. Thus 2500 trees ha^{-1} may be achieved with an approximate square pattern of 2.0×2.0 m, or an average of 1.7 m within rows and 2.4 m between them.

The effect on growth of a moderate degree of rectangularity is negligible but if distances between rows become excessive the canopy takes longer to close, hence the site is not fully used for a greater period and production is reduced. This has been well illustrated for extreme cases in a tropical alley cropping system of *Gliricidia sepium* (Karim and Savill 1991). Trees planted at 0.25×8.0 m and 0.5×4 m produced less than half the biomass per hectare at the time the canopy closed than trees planted at 1×2 m, though in all three cases the space available to each tree was 2 m^2.

Spacing and wood yield

Since spacing influences total production and the dimensions of forest produce, it can have a profound effect on the value of a crop. It is important to know how stand density influences production and how it can be manipulated. This has now been worked out for most major European plantation trees. Many countries have some form of production or yield models based on various spacing and thinning regimes which are normally entered via site index curves based on height and age.

In practice, there are two main aspects of spacing to consider—the effects of early or initial spacing and the effects of thinning. Initial spacings are quite often the densities at which plantations remain, except for mortality, especially where thinning is uneconomic or leads to windthrow. In Britain and Ireland, planting spacings recommended for conifers have decreased from about 4500 stems ha^{-1} in the 1930s to about 2500 stems ha^{-1} in the 1990s. The effects on productivity are shown in Fig. 10.2 for an experiment with *Picea sitchensis*, at 18.5 m top height. In it, spacings were within and well below this range. In this example densities greater than about 1500 stems ha^{-1} gave very little additional total production of stemwood to a minimum top diameter of 7 cm, but with fewer stems per hectare there was a rapid fall.

The fact that total production is not influenced by a very wide range of densities, nor as discussed later, by a wide range of thinning intensities, and the fact that dominant height is relatively independent of density led to the inference, known as Eichhorn's (1904) hypothesis (after the German who first propounded it) that total production from a stand, which is the volume currently standing plus anything removed in previous thinnings, is a function of its height (Assmann 1955; Pardé and Bouchon 1988). This is the basis of practically all yield models for forecasting production. Thus, by knowing the height of a stand, and within a wide range of stocking levels, it is possible to predict the total cumulative amount of timber produced since planting. This can be done with acceptable accuracy (usually to within 10–15 per cent) whether the stand has been thinned or not, and irrespective of age.

Total production is only one of the factors which influence value. The distribution of the volume in different size classes is often of much greater importance because the size and number of stems influence the cost of harvesting and the markets to which produce can be sold. Close spacings result in reduced total production of the larger sizes (Fig. 10.3) which are usually more valuable. Over long rotations, dense unthinned stands will give somewhat lower yields of usable timber than less dense ones due to mortality and higher levels of stand respiration, as discussed later. This is illustrated in Fig. 10.6 (p. 151).

Spacing and wood quality

The faster growth of individual trees at wider spacings, especially among many plantation conifers, causes an increase in the size of the core of juvenile wood

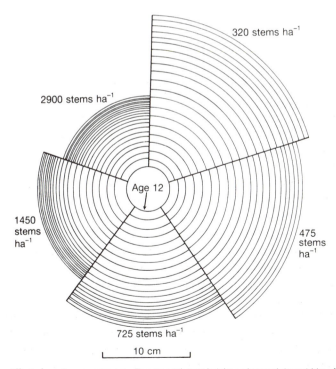

Fig. 10.3 Effect of spacing on mean tree diameter at breast height and annual ring widths of Sitka spruce (*Picea sitchensis*) 32 years after planting. (From Savill and Sandels 1983.)

which is undesirable for many end uses (e.g. Nepveu and Blachon 1989; Leban *et al.* 1991). Branches and hence knots are bigger and may be associated with more compression wood. A high wood density in conifers is usually associated with many desirable characteristics, such as suitability for end uses requiring strength. In some fast-growing species such as *Picea sitchensis* and *P. abies*, which have strength properties close to the lower limits for some industrial purposes, wide spacings may make the timber unacceptable for uses which command high prices (Brazier and Mobbs 1993). Though spacing and thinning both affect wood density in a stand, their influences are much less significant than the genetic variation which exists from one tree to another (Nepveu and Velling 1983; Colin *et al.* 1993). In both *Picea sitchensis* and *Pinus taeda* 60 to 70 per cent of the total variability is associated with tree-to-tree variation (Savill and Sandels 1983; Zobel *et al.* 1983). However, rapid growth in widely spaced trees results in wide annual growth rings and consequently in relatively wide bands of low-density earlywood being formed, which alternate with narrow bands of higher-density latewood. The resulting uneven texture may make the timber unsatisfactory for some purposes, such as joinery.

Other effects of spacing

Because spacing influences the time when canopy closes, weed suppression is likely to be faster in closely spaced stands. This reduces weeding costs and the fire hazard from flammable vegetation. Wider spacings require fewer plants which can save establishment costs. Spacing can also influence the period during which conditions for various animal and insect pests are favourable. Dense stands are more easily stressed during periods of drought and tend to be more susceptible to attacks by some insects and fungal diseases; wide spacing will result in reduced natural selection or less opportunity for managers to select, and so may lower the genetic quality of the final crop. Spacing can also influence the costs of pruning, thinning, and harvesting as shown in Table 10.1.

Thinning

Thinning is carried out to reduce stand density and hence to reduce competition; the remaining trees have more space to grow. It is also normally done to provide the owner with some revenue though, if this is not possible, as with some early thinnings, it is carried out in the expectation of greater returns later in the rotation. In broadleaved species the aim is usually to improve the quality of the final crop. There may also be less obvious reasons for thinning. For example Savill and Mather (1990) have shown that ring and star shakes, serious defects of oaks (*Quercus robur* and *Q. petraea*) on drought-prone sites, can be reduced if the most susceptible trees are removed in thinnings. They are trees with larger than average xylem vessels, and can be recognized and marked for removal at the time of leaf emergence in the spring because they flush later than other trees.

Table 10.1 Effect of wider spacing on costs of operations (+ = higher costs, − = lower costs) (modified from Evans 1992)

Operation	Effect on unit cost	Effect on period of operation	Overall effect	Comments
Ground preparation	−		−	Wider spaced furrows, fewer pits
Planting	−		−	Fewer plants needed, fewer to plant
Replacing dead trees	+		+	Higher survival % more important
Tending	−	+	None	More opportunity for mechanization but fewer trees to tend though usually for a longer time
Brashing			None	Fewer trees offset by thicker branches which take longer to cut
High pruning	+		+	Thicker branches
Fire protection		+	+	Delay in canopy closing lengthens period of risk from grass fires
Thinning	−		−	Fewer, larger trees can be removed to obtain similar volume
Harvesting (unthinned stands)	−		−	Fewer, larger trees to extract per cubic metre

Silvicultural condition of trees

For many descriptive purposes and especially in thinning, it is useful to describe the silvicultural condition of a tree in terms of canopy position and tree form. In even-aged plantations the four position and three form categories shown in Table 10.2 are usually adequate, though in irregular forests and for certain types of detailed work, more categories may be needed.

Responses of crops to thinning

Research into spacing, thinning, and primary production has indicated four main influences of thinning:

1. Provided a reasonably intact canopy is maintained, total production per unit area does not vary very much within a wide range of treatments. Production is spread over relatively few crop trees where thinning is frequent and heavy, meaning that they grow individually much more in volume than where the same increment per hectare has to be spread over many more trees (Hamilton 1976a, 1981).

Table 10.2 Description of silvicultural conditions of trees in even-aged plantations (adapted from Dawkins 1958, and Ford-Robertson 1971)

Position of crown in canopy	Definition
Dominant	Upper canopy trees with largely free-growing crowns exposed in entire vertical plan but usually in contact with others laterally.
Co-dominant	Upper canopy trees, similar to and often difficult to distinguish from dominants but crowns somewhat less free and more in contact with others.
Subdominant	Middle or lower canopy trees, partly exposed and partly shaded vertically by crowns, but leading shoots free. Also known as intermediate or dominated trees.
Suppressed	Lower canopy trees entirely shaded vertically by other crowns. Also known as subordinate or overtopped trees.
Stem and crown form	
Good	Of good size and development, straight bole, circular crown.
Tolerable	Just in the satisfactory class, stem may not be quite straight or crown asymmetrical, or sparse.
Defective	Unsound individual rendered useless by some defect such as a bent stem, large fork, heavy branches, permanent stagnation, or disease.

2. The removal of suppressed trees, which may often be on the verge of death, has a negligible effect on volume increment per unit area but if a proportion of the most vigorous trees are felled, there is an immediate reduction in growth because the remaining trees are not able to make full use of the resources of the site. Over a range of treatments, there is no permanent loss of increment. The stand not only recovers its normal level of growth, but also makes good the production lost when the increment was below normal immediately after thinning. The type of recovery pattern postulated by Bradley (1963) is illustrated in Fig. 10.4 and it can be seen that the length of the recovery cycle is proportional to the volume removed in thinning. The heavier the thinning, the longer the period of recovery. The reason for this pattern is not clearly understood but may be connected with the reduced level of stand respiration following the removal of some of the trees, making a greater net production possible for a period (see below).

3. If there is a large standing volume of timber and live branches, as in a stand which has not been thinned, it will consume more assimilated carbon for respiration than where lower volumes are maintained through thinning, hence net

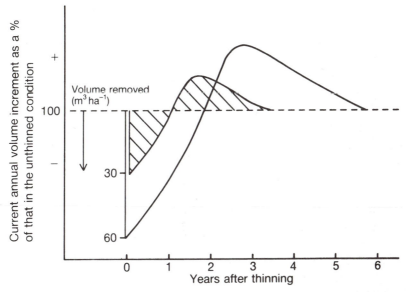

Fig. 10.4 Hypothetical relationship between volume per hectare removed in a thinning, and pattern of recovery of current annual increment. (Modified from Bradley 1963.)

production will be reduced in dense stands. The reduction will become more marked as the volume of standing timber increases.

4. There are two ways in which thinning practices can significantly reduce total cumulative production. The first is by consistently removing the most successful trees in the stand, the dominants, which are those most able to respond, and leaving the less efficient, smaller trees to grow. The second occurs if the intensity of thinning is so great that the site is not being efficiently used, when the crop cannot make up the lost production (Hamilton 1981). There may often be sound economic reasons for carrying out both these practices.

The precise effects of the interactions between the type, intensity, and cycle of thinning on total cumulative production per unit area have proved very difficult to determine with any precision over the range of treatments commonly used in plant-ation forestry. This is because differences are small and, for practical purposes, insignificant. The effects of these practices on the growth of individual trees are more clearly known and are of great importance to managers.

Thinning intensity

In an analysis of British thinning experiments by Bradley *et al.* (1966) it was shown that the proportion of growing stock which can safely be removed annually without decreasing future production is close to 70 per cent of the maximum mean annual increment over what is termed the normal thinning period. This period varies between species and between stands of different rates of growth within a species but occurs during the time when current annual increment is at its highest and ends shortly before the age of maximum mean annual increment. Annual thinning yields before and after the normal thinning period are lower. An example for a *Picea abies* stand with a maximum mean annual increment of 20 m^3 ha^{-1} year^{-1} is shown in Fig. 10.5. Here it is possible to remove 14 m^3 ha^{-1} year^{-1} (i.e. 70 per cent of 20) between the ages of 25 and 55 years without prejudicing future production. If a stand is thinned at this marginal intensity, by the end of the rotation about half the total cumulative production will have been removed in thinnings and the other half remains for the final harvest (Fig. 10.6). By contrast, in an unthinned stand of the same species, age, and growth rate, the standing volume will be about 10 per cent less than this cumulative total because competition will have caused some of the trees to die and hence some potentially useful production to be lost, as well as losses through increased respiration. In terms of basal area, 25 per cent or more must be removed per hectare before any substantial increases in diameter will occur to the remaining trees.

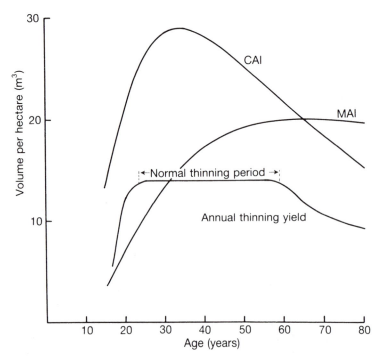

Fig. 10.5 Relationships between age, and mean annual increment (MAI), current annual increment (CAI), and annual thinning yields for yield class 20 m^3 ha^{-1} year^{-1} Norway spruce (*Picea abies*) thinned at the marginal intensity. (Data from Hamilton and Christie 1971.)

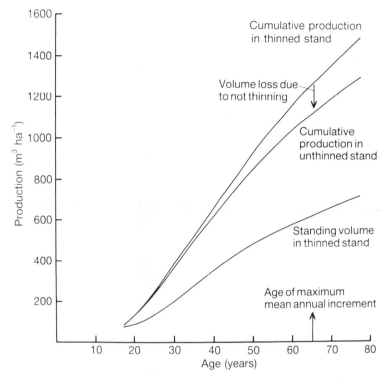

Fig. 10.6 Cumulative production in thinned and unthinned Norway spruce (*Picea abies*) of yield class 20 m³ ha⁻¹ year⁻¹ and planted at 2500 stems ha⁻¹. (Data from Edwards and Christie 1981.)

Time of first thinning and thinning cycle

The earliest a stand can be thinned without a loss in potential production varies according to the rate of growth in fully stocked stands. It can be as early as age 12 years for *Pinus pinaster* in south-west France, and 15 for *Pseudotsuga menziesii* in the Massif Central (e.g. Vannière 1984). In Britain, it varies between ages 20 to 35 for most coniferous species, and can be up to 50 years or more in the much harsher climate of Sweden. At this time in most regions the total standing volume per hectare varies between about 70 m³ ha⁻¹ for light-demanding species, such as *Larix*, to about 100 m³ ha⁻¹ for more shade-bearing ones.

The frequency or cycle of subsequent thinnings will depend to a large extent upon the economics of the enterprise. A stand could be thinned every year but it is normally more profitable to obtain the benefits of scale, by thinning occasionally and heavily rather than frequently and lightly. Cycles are commonly adjusted to remove 40 to 60 m³ ha⁻¹ in a single operation. Thus, a stand with a maximum mean annual increment of 10 m³ ha⁻¹ year⁻¹, thinned at the marginal intensity, can theoretically have 7 m³ ha⁻¹ removed every year. In practice it would probably be thinned on a 7- or 8-year cycle, to remove about 49 or 56 m^3 ha^{-1}. Long cycles are associated with slow-growing crops and short ones with more productive stands.

If the cycle is excessively long, very large gaps will be made in the canopy and the site will not be fully used for several years. Losses in cumulative production will then occur even though the marginal intensity has not been exceeded.

Type of thinning

A certain intensity of thinning can be achieved in many different ways regarding the kinds of trees that are removed. A useful indicator of the size of tree removed, and hence for describing the thinning type, is the ratio of the mean volume of thinnings (v) to the mean volume of the stand *before* thinning (V). The type of thinning normally carried out may broadly be classified as follows:

1. Systematic methods
    Line and strip thinnings—v/V ratio 1.0

2. Selective methods
    Low thinning— v/V ratio about 0.6
    Intermediate thinning— v/V ratio about 0.8
    Crown, or high thinning— v/V ratio about 1.2

3. Combinations of 1 and 2
    e.g. Queensland selection thinning— v/V ratio variable

Systematic or neutral methods

Systematic thinning entails the removal of trees in a predetermined way, regardless of their individual size or quality. Complete lines of trees are usually removed or strips which do not follow planting rows, though other systematic patterns are possible. Systematic methods are neutral in terms of the size of trees removed. On average they will be the same as the average in the stand as a whole, hence $v/V = 1$. The various types, techniques, and responses of crops to systematic thinning have been described by Hamilton (1976b, 1980).

Systematic methods are now very commonly employed as first thinnings which, because of the small average tree size, usually yield little income. Neither brashing nor selective marking of trees is necessary so costs are much lower. Also, extraction routes for the first and subsequent thinnings are created (Fig. 10.7).

The disadvantages are that defective trees are not deliberately removed, and the removal of a complete range of tree sizes, including some dominants, results in a loss of total cumulative production, often amounting to 6 to 8 m^3 ha^{-1} (Hamilton 1976b). If line thinning is done only once in the life of a crop, such losses are insignificant in terms of total production, amounting to only about 1 per cent. However, even one line thinning can lead to the development, through larger branches, of some reaction wood on the exposed trees along the edges of lines. It

Fig. 10.7 This stand of *Pinus sylvestris* in Finland has been systematically thinned, by removing lines of trees, and is being high pruned.

may also expose a stand to serious risks of windthrow. For example, Oswald and Pardé (1976) and Oswald (1980) clearly demonstrated the advantage of selective over systematic thinning for resistance to various climatic hazards—particularly windthrow—in experimental *Pseudotsuga menziesii* and *Abies grandis* plantations.

Selective thinning methods

Low thinning is a selective thinning that particularly favours the dominants by removing trees progressively from the lower to the upper canopy. The first trees to be marked for removal are suppressed trees, then subdominants ($v/V = 0.6$); in heavy, low thinnings some of the dominants may then have to be felled to achieve the desired volume. At the other extreme, crown thinning favours the most promising stems which are usually, but not necessarily, dominants. Theoretically, trees from any canopy class which are interfering with selected trees may be removed. Most commonly, poorer quality dominants and codominants are usually removed and these have higher average volumes than in other types of thinning ($v/V =$ about 1.2).

The commonest type of selective thinning is intermediate between low and crown thinning ($v/V = 0.8$), but in all selective thinnings defective trees, especially defective dominants, are also removed. Included among numerous variations to these methods is the early selection of final crop trees in broadleaved stands, as now widely practised in Britain and France (Lanier 1986; Evans 1982*a*). The best trees are marked when they are young and favoured in subsequent thinnings. Because some inevitably become damaged or do not grow as well as expected, it is necessary to mark at the outset two or three times the number that will actually form the final crop.

The effects of the various types of thinning on crop productivity have been discussed in detail by Assmann (1970), and in the context of a classical British trial, the Bowmont Thinning Experiment of *Picea abies* in Scotland, by Hamilton (1976*a*). At Bowmont, four thinning intensities were applied, described as:

B grade—involving the removal of mainly dead and dying trees only, an extreme form of low thinning;

C grade—a moderate low thinning, intermediate between B and D;

D grade—a heavy low thinning in which only the best trees, mainly dominants, were left and given space for full crown development. It was somewhat heavier than the marginal thinning intensity;

LC grade—a moderate crown thinning, resembling D except that suppressed and subdominant trees were left to fill spaces between the best dominants. Defective dominants were removed.

The results (Table 10.3), 64 years after planting, confirm the expected trends of larger individual trees being found in the heavier thinning treatments, but with little overall difference in total production except in treatment B.

The significantly lower production in treatment B may partly be a result of a severe infestation of *Heterobasidion annosum* which existed. In *Picea abies* and *Pinus sylvestris* (and presumably other conifers too), high density stands are more likely to be infected than those of low density (Belyi 1975; Tribun *et al.* 1983). It is

Table 10.3 Some results of the Bowmont Thinning Experiment (in *Picea abies*) 64 years after planting (from Hamilton 1976*a*)

Treatment	Stems ha⁻¹ (no.)	Top ht (m)	Mean tree		Vol. ha⁻¹) standing (m³)	Cumulative crop yield to date	
			dbh* (cm)	Vol. (m³)		Basal area ha⁻¹ (m²)	Vol. ha⁻¹ (m³)
B	2204	19.6	21.7	0.333	733	109	861
C	1062	20.2	27.8	0.569	605	123	965
D	327	21.6	42.5	1.387	453	127	953
LC	691	20.9	30.2	0.641	443	128	924

*dbh = diameter at breast height.

also probably partly due to higher levels of respiration caused by the much greater volume of standing timber (p. 148).

Combinations of systematic and selective thinning methods

Where circumstances permit, there are attractions in combining some of the cost savings of purely systematic thinning with the ability to select for vigour and good growth form. The Queensland selection method (Forestry Department, Brisbane 1963) is a combined method and was developed for treating hoop pine (*Araucaria cunninghamii*) plantations in Queensland and has a wider potential for use in temperate forests.

It involves selecting dominant trees along rows for retention and marking them in an obvious way, in Queensland by high pruning. The trees are selected by considering running groups of four trees (counting blanks as trees) on a row by row basis as, for example, in Table 10.4. Thus, the first group comprises tree numbers 1, 2, 3, and 4. Of these, 1 is the best stem and is selected (S) and pruned. The next group is tree numbers 2, 3, 4, and 5, and of these 5 is selected. In the next group, 3, 4, 5, and 6, tree 6 is better than 5 and therefore the best in the group and is selected. In the next group, 4, 5, 6, and 7, tree 6 remains the best of this new group, so no additional tree is selected, and so on.

Selection can be done quickly by unskilled workers and results in about 40 per cent of the stems originally planted being selected, though the percentage can be varied by altering the size of the group. Thinning rules can be applied equally simply in relation to the selected trees. For example, all trees which are equal in height to or taller than the selected trees and adjacent to them may be removed in the first thinning.

Table 10.4 An example of the selection of trees in Queensland high pruning

Selected	S				S	S	S				S	
Tree No.	1	2	3	4	5	6	7	8	9	10	11	12

Pre-commercial thinning

Thinning stands before trees are big enough to be sold is known variously as pre-commercial thinning, respacing, thinning-to-waste, or unmerchantable thinning. It is often carried out at about the time canopy closes while access is still easy. The thinned trees may be killed chemically, for example by an injection of glyphosate, or felled and left in the stand.

In planted crops foresters have the options either of planting at the density required at the time of clear-felling or the first commercial thinning, or of planting more densely and carrying out a precommercial thinning. The arguments in favour of close planting and precommercial thinning have been discussed by Edwards

(1980) and Savill and Sandels (1983) and include more certain establishment, the opportunity for selecting the most vigorous trees, smaller branches and faster self-pruning, and, above all, a narrow juvenile core. The disadvantages are mainly those of cost: establishment is more costly and the precommercial thinning itself is expensive.

Precommercial thinning is attractive when there are difficulties in selling of small-sized material profitably. It is also seen as one way of reducing density before crops reach a height at which they become susceptible to windthrow. It is also occasionally done to encourage naturally regenerating broadleaved trees, to create diversity in otherwise coniferous monocultures.

Pruning

Pruning may be done for one of three reasons:

1. Complete or partial low pruning (called brashing) to head height for access to stands, is to make working easier and safer and sometimes for lowering susceptibility to fire.

2. To improve the quality of timber by reducing the extent of knots on the potentially most valuable lower part of the stem. This is usually done to a maximum height of about 6 m (Fig. 10.7).

3. Many broadleaved trees need what is called formative pruning to prevent forking and heavy branching.

The persistence of branches on trees varies with species and spacing. In similar conditions, tolerant trees retain living branches for longer than intolerant species. Among the latter, the duration of retention of dead branches also varies considerably; *Larix*, *Fraxinus excelsior,* and *Eucalyptus* self-prune well, *Picea* spp. and *Abies* moderately well, while *Pinus*, *Prunus avium*, and *Populus* spp. do not (Boudru 1986,1989). Persistent dead branches can lead to the formation of loose or decayed knots. To obtain the economic advantages of clear, pruned timber, the amount of knotty wood must be kept to a small core in the centre of the stem, usually no more than 10 to 15 cm diameter.

Effects of pruning on tree growth

Pruning stems as small as 10 to 15 cm diameter necessitates cutting off living branches. If too many are removed in normal plantation conditions, the photosynthetic capacity of the tree is reduced and growth suffers. It is, however, possible to remove some branches from the lower crowns of trees without affecting growth at all because, in low levels of light, leaves on these branches may respire as much or conceivably more than they photosynthesize, and so contribute little or they may even be a drain on the resources of the tree. Wang *et al.* (1980), for example, found that removing 10 per cent or slightly more of the live crown of *Cryptomeria japonica* improved growth.

From a detailed analysis of European and north American pruning experiments, largely on conifers, Møller (1960) concluded that the removal of up to 25 per cent of the live crown has no effect on the growth of most species. Removal of one-third of the crown reduces production by only a fraction of 1 per cent by the end of the rotation and in some species half or more can safely be removed. More severe pruning reduces both height growth, and even more markedly, diameter growth. Some species are more sensitive than others to increasing intensities of pruning. For example, *Pseudotsuga menziesii* is more sensitive than *Picea* which, in turn, is more sensitive than *Pinus* (Henman 1963). Excessive green pruning may sometimes lead to stress and all its attendant dangers of increased risks of attack by insects and pathogens.

Pruning young trees in very widely spaced stands may result in somewhat different responses. Funk (1979) found in 5-year-old black walnut (*Juglans nigra*) planted at 6.1 × 6.1 m that height growth increased over 3 years with increasing intensities of pruning (which ranged from no pruning to 80 per cent of the total length of the stem). Though volume growth did not differ among the various pruning intensities, increment was greater further up the stem in the more heavily pruned trees, making them more cylindrical.

It is usually perfectly safe to remove live branches from trees without any undue risk of infection and decay, provided they are small and contain no heartwood. This is especially the case with conifers because, when trees are injured by pruning, or in any other way, wounds inevitably become infected but trees are able to compartmentalize or wall-off the infected wound with protective barrier zones, confining it within the diameter of the tree at the time of wounding. Thus, if a tree is pruned when it is 10 cm in diameter, no more than a 10-cm core of defect may develop.

Pruning and wood quality

Apart from eliminating knots, pruning has other desirable effects on wood quality in some species. The removal of live branches reduces the formation of low-density juvenile wood in the pruned part of the stem, as well as preventing the formation of reaction wood around branches (Keller and Thiercelin 1984). In *Pinus radiata* it leads to a temporary increase in wood density, by up to 7 per cent, for 2 or 3 years after treatment as a result of a greater proportion of latewood being formed. There may also be increases in fibre length and decreases in spirality of the grain (Gerischer and de Villiers 1963). In some broadleaved trees, it is believed that early pruning may also confine ring shake and discoloration to a small central core (Butin and Shigo 1981). Among the negative effects of pruning are tendencies for resin pockets to develop and bark to become enclosed around the pruning scar. Epicormic branches can become a serious nuisance in some species such as *Quercus*, especially if the stems have been heavily pruned.

Pruning practice

Where pruning is carried out it is normally done in two or more stages or lifts (Fig. 10.8), to achieve about 6 m of clear stem. Pruning in stages is necessary if the pruned core is to be kept small enough while allowing sufficient crown to be retained to maintain vigour. The first pruning often occurs when the breast height diameters of the best trees in a stand average 12 cm, and are no more than 15 cm. The stems are pruned up to about 4 m in height, where the diameters are about 10 cm. The second occurs a few years later when the diameters at the 4 m level have increased to about 12 cm, and pruning continues up to 6 m in height or where the stem diameter is 10 cm. Only rarely, such as with *Populus* spp., would a third or fourth pruning to 8 m be carried out. It is difficult to manipulate even the longest pruning tools from ground level to heights much above 6 m, so if greater pruned lengths are desired ladders or other equipment are needed which greatly increase the expense. It is seldom considered worth pruning unless the top diameter of the pruned section grows to at least 40 cm diameter over bark at the time of felling. Hence only the number of trees likely to grow to this size plus possibly 50 per cent as an insurance, the final crop trees, are normally pruned. If selected trees are pruned heavily in dense stands, it will also be necessary to thin round them otherwise their dominance may be lost, so pruning is often accompanied by a thinning. Detailed pruning schedules have been produced in some countries (e.g. Hubert and Courraud 1987 for France).

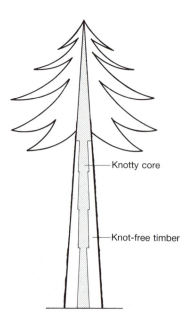

Fig. 10.8 Diagrammatic longitudinal section of a tree which has received four prunings at intervals of a few years. (From Henman 1963.)

There does not appear to be a clear consensus about the precise time of year at which trees should be pruned. For many species, the period during which wounds are particularly susceptible to infection should be avoided, so from this point of view it is best to avoid the dormant season. For some species, including *Prunus* and *Fagus*, summer pruning is recommended (e.g. Kerr and Evans 1993). Periods of intense cold and severe drought should be avoided, as should, in susceptible species, periods when epicormic shoots may sprout (hence December is normally the recommended time for pruning oak).

Apart from broadleaved species, where formative pruning may be essential if a stand of any value is to be obtained, practices vary widely. Pruning is an expensive operation. In parts of Germany, France, Belgium, and other European countries, prices paid for pruned timber can be up to twice as high as unpruned (e.g. Luscher *et al.* 1991). Official certification schemes for pruned stands are sometimes applied as labels of quality (e.g. Maugé 1986,1987; Hugon 1994). In other countries such as Great Britain and Ireland, there is less tradition of growing high quality timber, and so pruning is relatively rare. It is never worthwhile economically to prune trees that will be used for low-quality purposes.

Rotation—clear-felling

The term rotation is used to describe the interval between successive crop regenerations. Rotations can be determined for technical reasons, for example growing to a specified log size, or for silvicultural reasons, or for maximizing volume yields or financial returns. For some private woodland owners they may be determined by the need to realize capital or aid cash flow. Sometimes rotations are determined by force of habit or circumstances and quite often they remain undecided. However, in that plantation forests are usually established to fulfil a real or anticipated need, the tendency is usually towards short rotations (Fenton 1967).

The nineteenth-century concept of a 'normal forest' embodied the idea of an unvarying rotation (see p. 55). Today a fixed rotation could well be an obstacle to progress and the choice of felling time is a subject for constant review, according to changes in the objectives of an enterprise or the circumstances which influence the method of attaining current objectives (Johnston *et al.* 1967). Management today is also more flexible because of opportunities for stand improvement, achieved through added fertilizers, tree breeding, and other silvicultural measures.

Furthermore, the best time for felling a stand is usually considered in the light of what the whole enterprise has available and rotations are often varied from the best theoretical time so as to provide a measure of continuity of supply and of employment. It would, for example, clearly be unwise to fell large adjacent areas of woodland judged as overmature by current criteria as this might depress prices and have far-reaching adverse amenity and conservation effects. In other circumstances, it might be prudent to fell some crops early to keep a new industry supplied. In many cases, when rotations are clearly laid down, they are simply rationing devices for allocating resources to wood-using industries. Occasionally, there are biological or silvicultural grounds for clear-felling at a particular time. For example, high wind-throw hazard areas may be felled prematurely before windthrow sets in, and to achieve natural regeneration it may be necessary to fell during the period in a stand's life when it bears heavy seed crops, and in a particular year when seed has been produced. In a few countries (e.g. Switzerland, Slovenia) there are legal restrictions to clear-felling.

Large forestry companies and state forest services often consider the best time to fell is when stands yield either maximum net discounted revenue or some other measure of profitability. With many conifers maximum net discounted revenue often happens to coincide roughly with rotations of maximum mean annual incre-ment though, depending on the region, they may be shorter (as in Britain) or longer (as in Sweden) by 10 to 15 years. For broadleaved species like *Quercus*, much longer rotations are usually necessary to take advantage of the sharp rise in value, as trees develop from the firewood, mining, and fencing sizes to planking quality sawlogs, or for the very best trees, veneer.

In some countries quite different financial criteria are often used. For example long rotations to yield maximum value at the time of felling are favoured where high quality timber can be assured. Maximizing net discounted revenue favours short rotations and consequently rather low-quality, fast-grown timber. Thus, in northern Germany, for example, high-quality valuable *Pinus sylvestris* is often grown on a 200-year rotation for joinery timber. On similar sites, with similar growth rates in Britain, it is seen as more profitable to grow three rotations of *Pinus sylvestris* in the same period, or preferably four of *Pinus nigra* subsp. *laricio*, even though the value of timber per cubic metre at the end of each is considerably less.

Spacing, thinning, and pruning practice

In the words of Craib (1939), the purpose of most commercial forestry, and hence of manipulating spacing, pruning, and determining the length of rotations is 'the production of neither the greatest volume, nor the largest sizes in a given time, but the production of material which will most suitably fit the demands of the market, yielding, at the same time, the greatest revenue'.

Unfortunately, this is much more easily said than done. Many forest owners have a general idea of what they are trying to achieve but no one can know pre-cisely what market demands will be decades ahead. Nor do they know the precise interdependence between growth and quality, and the effects of stand age, site, and the costs of spacing, thinning, and pruning. The analysis and integration of numer-ous experiments is needed.

Practices differ very widely among countries. In some parts of Germany, *Pinus sylvestris* is planted at 10 000 to 18 000 trees ha^{-1}, 'a considerable reduction from earlier practice' (Otto 1976). Stands are thinned and pruned regularly and felled after rotations of about 200 years. In Britain, most conifers are planted at densities of about 2500 stems ha^{-1}. They are not always thinned because of risks of wind-throw; pruning is seldom carried out and rotations are seldom longer than 70 years. At the other extreme, in New Zealand, where there is a surplus of plantation-grown timber for home demand the main emphasis is upon producing high-quality, knot-free timber for export. On some land, *Pinus radiata* is planted at densities of only about 200 stems ha^{-1}. Scrupulous attention is paid to pruning. Since the canopy does not close until the end of the rotation, the grass growing among the trees is grazed or used for silage. Rotations are likely to be about 28 years with this regime, having been reduced from 40 years. It is clear that Craib was correct but regimes which achieve the best results differ widely, according to the economic and other conditions in a country and often according to the particular circumstances of individual owners as well.

11 Protecting plantations from damage by wildlife, pests, and pathogens

A plantation represents a considerable investment and so concern for its health is an important part of silviculture, as discussed in Chapter 2. In natural forests seedlings and mature trees are essential links in the ecosystems' food chain. Many organisms are partly or fully dependent on trees for their existence.

Though damage to plantations is often unpredictable, irregular, and very variable in its severity, fortunately most plantation schemes in temperate and boreal regions are successful because serious damage is rare and adequate protection is usually possible, though sometimes expensive.

Predictability of damage

To prevent damage, a distinction must be made between newly planted areas and those being reforested. Different approaches often have to be taken to them. In the case of newly planted sites, many animal pests and fungal pathogens are often absent. With reforestation most organisms are already present and able to respond to a new food supply or a suitable substrate on which to live.

Some types of damage are predictable. For example most foresters can anticipate that coniferous seedlings on newly replanted sites are likely to be attacked by pine weevils (*Hylobius abietus*), an insect which breeds under the bark of coniferous stumps in clear-cut areas. In some regions, no young trees will survive browsing from deer without fencing. Stem and root rots commonly reduce revenue from a conifer plantation if the stumps are not treated to prevent infection.

Other types of damage are unpredictable, irregular, and very variable in severity. Climatic extremes can play an important part in making a stand sensitive to attacks by pests. A correct choice of species and provenance ensures that most plantation schemes are successful, but exceptions do occur.

The structure of a plantation, normally a monoculture with regularly spaced, even-aged trees, makes it more susceptible to insect pests and pathogens than a natural forest where many of their predators and competitors also occur. But with good site preparation and maintenance, plantation trees are less likely to suffer from competition, badly aerated soil or to grow on poor microsites. Hence the risks of damage from insects and weakly pathogenic organisms are often less than in natural forests.

Many animal pests and fungi specialize on particular species and sizes of trees. Some, including the larvae of various butterflies, moths, and hymenoptera, thrive on healthy vigorous trees. Elk, deer, wild boars, rabbits, voles, and capercaillie all prefer fertilized, fast-growing trees. Others, including bark beetles and several normally saprophytic fungi, such as *Armillaria* spp., attack stressed trees.

Stress

Many potential pests can live in the forest at population levels that are so low as to be harmless. When the trees become stressed, attacks are facilitated and an epidemic outbreak can be very damaging. Stresses to planted or natural forests can be expected in the following conditions:

◆ extremes of temperature;

◆ deficits or excesses of water;

◆ high levels of radiation;

◆ chemical stresses induced by a deficiency of nutrients or excess of salts or toxic gases;

◆ windy environments;

◆ attacks by other pests, such as defoliators and/or leaf diseases.

Some plants have adaptations which enable them to avoid or to tolerate stresses (the latter is termed 'hardiness' by the layman) and these are discussed in detail by Levitt (1971). The absence of an appropriate mechanism for tolerance or avoidance can upset physiological balances in a tree and this leads to conditions which impair vigour. Defence mechanisms are then lowered, disorders induced, and pest invasions are more likely.

In unstressed, natural conditions damaging pathogens are most prevalent at stages of development of hosts where they do not greatly affect reproductive vigour, especially seedling and senescent phases. Forest nursery plants tend, therefore, to need more protection than older crops. The most prevalent disease fungi of adult reproductive phases are specialized species which pass greater or lesser periods of symbiotic existence with their hosts and allow them considerable physiological and reproductive vigour (Harley 1971). These species are non-aggressive pathogens. They are commonly found as saprophytes on dead and dying plant tissues and do no damage as long as host vigour is high. Only if vigour is reduced are they able to function as pathogens. The same is true of many insects, which do little damage to healthy trees but can become very harmful to stressed ones.

Anything that reduces sap pressure in trees tends to predispose them to attack by shoot-, bark-, and wood-boring insects.

It is important to know which stresses predispose trees to damage by particular pests, but the sites most liable to stresses include reclaimed sites (Chapter 14), polluted sites, deep peats, drought-prone sands, and forests at high altitude.

Maintaining a high level of vigour by planting species which tolerate stress is still the most effective means of preventing diseases induced by environmental extremes (Schoenweiss 1981). This, more than anything, emphasizes the need for careful selection of species and provenance for the site in question.

Stress cannot always be avoided since the physical environment is never constant. During the life of a plantation, years with unusually long droughts, low winter temperatures, or severe gales inevitably occur and create the conditions for weakly pathogenic organisms to cause damage: some examples of these are given in Chapter 2.

An understanding of the factors which cause shifts in the size and structure of pest populations can often indicate the best means of protection. Great advances have been made in these respects by theoretical ecologists. Southwood (1981) and Conway (1981) have discussed the relevance of the bionomic characters of organisms, such as size, longevity, fecundity, and mobility, that have developed to maximize their fitness to their environments. These factors cause selection along the so-called $r–K$ continuum. The term r refers to the rate of increase in numbers of a population and K to the maximum sustainable number of individuals of that species in a habitat, the carrying capacity.

r- and *K*-strategies

Extreme r-strategists, whether pests or not, continually colonize and exploit habitats of an ephemeral nature or respond to variable or unpredictable environmental conditions with an opportunistic 'boom and bust' strategy. Fecundity is high (hence r) and the generation time, or juvenile stage, is short, but as the habitats are virtual ecological vacuums, a high competitive ability is not required. Migration or dispersal, often of a very wasteful nature, is a major component of the process, so they turn up quickly and breed rapidly in suitable habitats. Some r-selected species produce high numbers of sexually derived progeny which ensures a high probability of an appropriate genotype landing on a safe site: others, such as aphids, reproduce asexually. Populations are unstable and fall rapidly as conditions deteriorate. In the case of insects, the fall is often caused by rapid dispersal rather than death or reduced levels of reproduction. The main defences against predators include mobility and a measure of synchrony in reproduction or germination which temporarily satiates predators.

Most of the classic pests, such as rats, locusts, and rusts of wheat, exhibit typical r-strategies. The widespread use of monocultures in agriculture and plantation forestry has greatly increased the availability of favourable habitats for r-pests and they include most of the more important pests of foliage and roots. r-pests are difficult to control because of the massive, damaging outbreaks caused by their rapid rises in numbers and high dispersal ability.

K-strategists, by contrast, are specialists which occupy narrow niches in stable crowded habitats. They evolve towards maintaining populations closely to equilibrium levels. Fecundity, or in the case of trees recruitment, is not high but K-strategists are competitive, typically large in size, with a long period between generations. Survival is good through investment in complex defence mechanisms. There is less tendency towards migration than in r-strategists. For the population to persist, the habitat must not be irrevocably disturbed. If mortality does occur, populations need to return quickly to equilibrium levels through a rapid increase in the reproduction rate or competitors may seize the resources: K-strategists are not well adapted to recover from population densities significantly below their equilibrium levels and if depressed to such low levels, they may become extinct.

In natural forests, K-strategists seldom become pests and, in fact, more often need the concern of conservationists, but in plantations if they become pests, they can be both persistent and troublesome.

The great majority of pests, and especially forest pests, lie between the extremes of the r–K continuum. Their most important characteristic is the degree to which numbers are regulated by their own intraspecific mechanisms of competition for food, space, or mates. They are also regulated by natural predators which are normally present with them in the ecosystem, or by the inbuilt defence mechanisms of trees. This regulation is often sufficient to keep the damage caused to crops below the level at which control is required. They therefore become pests only occasionally, possibly due to stress caused by climatic irregularities.

The best known intermediate pests are those which have been imported to new regions of the world but their natural predators are left in the country of origin (see Chapter 2). Others become important because man has eliminated or reduced the efficiency of their enemies, including hitherto insignificant pests whose natural predators are killed with pesticides. Man may also have increased the food supply and altered the abiotic environment and hence provided conditions which allow reproduction to increase with reduced competition.

To avoid damage to a plantation, it is wise to estimate the risks of attack by pests. It is almost always cheaper to prepare the site and to select species and sizes of plants that will withstand possible attacks than to repair the damage by replanting or or at least replacing many dead plants.

Practical considerations

In natural forests, most organisms have survived together for millennia. Each tree species has evolved a strategy for survival and expansion at the expense of other organisms. When establishing a plantation it is useful to try to understand as much as possible about the relations between the tree and the associated organisms in the habitat. For example many trees have developed defences against browsing including unpalatability and thorniness; thick bark and high resin flows provide some protection against insects and fungi. Some species respond to attack by producing more seed.

Knowledge of the complex interactions between trees species and their potential pests may suggest appropriate protection or preventative measures. A correct identification of the primary cause of damage is critical for control.

Identification of pest and pathogens

It is normally easy to associate characteristic browsing and debarking patterns on trees with particular animals from the size and marks of teeth. Fungal and insect attacks are less visible and consequently more difficult to identify, even though the galleries of some bark beetles, for example, are very characteristic. The primary attack is often followed by more visible symptoms of secondary agents.

A potentially serious problem can easily pass unnoticed for a long time. Some butt rots, for example *Heterobasidion annosum* and *Phaeolus schweinitzii*, show scarcely any visible symptoms of infection until sporophores are present. Reductions in the quantities of protective chemicals in the foliage of stressed trees, such as resins and tannins, are not visible yet such reductions may make the trees more palatable to many insects. Even where there is visible damage, it may easily be overlooked for years. This happened in Britain with the great spruce bark beetle, *Dendroctonus micans*, a serious, potential hazard native to continental Europe. It was first noticed in England and Wales in 1982 but evidence from the callusing of damaged trees indicated that it had already been in the country for about 9 years.

Normally, problems are first detected by observation: looking for fruiting bodies of fungi, damaged foliage, fruits or shoots, or more general symptoms such as poor growth, dieback, beetle emergence holes, or discoloured or sparse foliage. Attacks by many species of both fungi and insects are likely to occur at particular stages in the rotation and so can be anticipated to some extent (Murray 1979; Crooke 1979). Severe outbreaks of some pests can sometimes be predicted from recent climatic patterns and certain sites can be classified as high-risk areas from previous experience.

Once damage to a crop has been detected, the correct identification of the primary cause is essential. This can be time-consuming and difficult and is usually the province of specialized forest pathologists and entomologists. Failure to identify the primary cause and treating a secondary infection or attack can lead to a very wasteful use of resources.

It may be necessary to spend time, often years in the case of a 'new' pest, studying its life history, ecology, and the factors which predispose trees to attack, before a wholly appropriate strategy for control can be devised.

Preventative measures

Good silviculture takes potential pest damage to the plantation into consideration for the whole rotation. It is always cheaper to foresee a threat than to control the agent after the outbreak or to fell and replant the forest.

Protection against the import of exotic pests and diseases

European governments have for long enacted laws to prevent the uncontrolled movement of animals and plants between countries. The reason for this is that many organisms are surprisingly often native to quite small regions. Experience with moving them over long distances into new ecosystems has sometimes been beneficial and occasionally catastrophic. One reason for the good health and high productivity of many exotic trees that have been introduced to Europe, primarily from North America, is believed to be the absence of specialized pathogens (see Chapter 2).

Unfortunate examples for commercial forestry include the introduction of browsing and bark-stripping animals to new regions: rabbits to Sweden and elsewhere, grey squirrels (Fig. 11.1) to Britain, and various deer to Britain and New Zealand. More serious examples are found among introductions of parasitic fungi and insects. A classical example is the introduction of the white pine blister rust, *Cronartium ribicola*, to North America (see Karlman 1981 and literature cited therein). This fungus is a weak and harmless parasite on indigenous five-needled pines (*Pinus cembra* and *P. sibirica*) in Europe. The American *Pinus strobus* was introduced into France and Germany on a large scale early in the nineteenth century. In the middle of that century the plantations were suddenly attacked by the rust. The epidemic spread over the entire continent during the course of only 30 years and the effect was catastrophic since *P. strobus* possessed no resistance. *C. ribicola* was introduced to north-eastern United States in an export of *P. strobus* seedlings from Germany between 1906 and 1910. From this, it successfully spread throughout the natural range of *P. strobus*. Another export of plants to Vancouver caused immense damage to other five-needled pines: *P. monticola, P. lambertiana, and P. flexilis*. Progress in research towards the development of resistance in these species has been slow (Ziller 1979), though the most promising means of control has been through selection and breeding of individual resistant trees (Agrios 1978).

Another example, given by Karlman (1981), is the world-wide spread of Dutch elm disease, *Ophiostoma ulmi* and *O. novo-ulmi*. This insect-borne parasitic fungus was first discovered in France in 1918. The disease was noticed in England in 1927, and shortly after, it was carried over the Atlantic to the United States, causing severe damage to *Ulmus* spp. in North America. The disease crossed the Atlantic again in the 1960s in a much more virulent form and soon eradicated most of the adult *Ulmus* spp. in western Europe.

Such examples, and others from horticulture and agriculture, have persuaded most governments to legislate to control the transfer plant and animal material

Fig. 11.1 Damage caused by grey squirrels debarking the bases of a young beech tree (*Fagus sylvatica*) in England. This exotic pest is, fortunately, still very rare in continental Europe but is a very serious problem in Britain.

between countries. Customs officials are responsible for enforcing this part of the forest hygiene programme through inspection at borders and enforcing quarantine for some animals. From the forest hygiene point of view, it is wise to adhere to these regulations.

Within-country protection from pests

Most European countries regulate forest practices to prevent pest epidemics, and regulations are enforced through special agencies which also advise and assist owners when a pest outbreak is expected or is in progress. Examples of situations where the risk of an epidemic is great are when a population of bark beetles increases faster than the predators after storm damage, fire, or a severe drought. The storage of unbarked wood (Fig. 11.2) during the breeding period of the insects can also result in epidemics.

Fig. 11.2 In much of Europe, felled trees must be debarked to eliminate potential breeding sites for the bark beetle *Tomicus piniperda*, unless the logs are sawn almost immediately.

Prevention of damage by large mammals

In Europe, many of the predators of large mammals, such as moose, deer, and wild boars, which partly feed on or damage trees, are now extinct. To protect plantations from severe damage, hunting to reduce numbers is important (it also provides recreation and meat). Hunters and foresters have to agree upon sustainable population densities for each species so that damage to the forest is not excessive.

Fencing plantations against deer, wild boars, hares, and rabbits is expensive but sometimes essential. For a fence to be effective, maintenance is required, and normally some hunting inside the enclosure to eliminate potentially damaging animals, for up to 10 years. Wire netting is most commonly used but electric fences are now cheap, though they require almost daily attention.

Individual plant protection, in the form of 1 m or taller plastic tubes, has become more common recently (see Fig. 7.3, p. 103). These tubes provide effective protection against browsing deer, grazing animals (e.g. in silvopastoral systems—see Chapter 16) and attacks from voles, especially to broadleaved trees. Though they are expensive, in some circumstances they offer the only means of success. The tubes are, unfortunately, lethal to some insectivorous birds such as stonechats (*Saxicola torquata*) if they get into them.

A method for reducing damage to seedlings by deer, hares, and rabbits is to apply repellents—non-toxic chemicals with unpleasant smells or tastes—either to the site or to the trees. Most are effective only for short periods and then have to be reapplied. They are consequently very expensive to use unless only small numbers of trees have to be protected (Pepper 1978).

When using poisons to prevent severe attacks to plantations it is important to choose chemicals, or to apply them in such a way, that only the target pests are affected. For example warfarin-treated grain is used to control voles in Sweden, and it is applied in long, narrow tubes that voles can enter but not birds. A similar approach is taken to controlling grey squirrels in Britain.

Preventative protection against fungi

The fungus that causes the most serious economic losses in Europe, and the only one against which control measures are routinely applied, is the butt and root rot fungus, *Heterobasidion annosum* (Fig. 11.3). For example it is the main cause of decay in 50- to 60-year-old *Pinus sylvestris* in Thetford forest in the east of England. On some sites, up to 15 per cent of the trees suffer from the rot, with decay occasionally extending 2 m or more up the stem (Greig 1995). It attacks most species of conifers, and gains entry to first rotation stands through airborne

Fig.11.3 Damage caused by *Heterobasidion annosum*—the dark, central parts of the stem and stump are decayed. (Photo: M. Morelet, INRA.)

basidiospores infecting freshly cut stumps after thinning. After germination, the mycelium grows down into the stump and its root system from which, through root contacts, it infects neighbouring trees, spreading into the heartwood of the stem and causing a butt rot. Less frequently, and normally only in young trees, it may attack the whole root system and cause the trees to die. In second rotation crops, infection occurs much earlier, originating from the stumps of the previous crop. Preventative measures include thinning during periods of intense winter cold (in northern Europe), or spraying the stump with a suspension of spores of a competitive fungus (e.g. *Peniophora gigantea* is used on pines in eastern England), or with urea, or other chemicals which inhibit the growth of the *Heterobasidion*. On very badly affected sites, such as parts of Thetford Forest, the removal of stumps at the end of the rotation may be the only effective remedy (see Fig. 5.4, p. 72).

Many fungi do not cause epidemics but can result in reductions in quality and losses in production of infected trees. Among these are many of the rust fungi in the genus *Melampsora* that primarily attack leaves and consequently reduce photosynthetic production. These rusts normally alternate their hosts with different species for different parts of their life cycles. The presence of both host species on a site substantially increases the risks of infection. For example, *Melampsora pinitorqua* alternates between aspen (*Populus tremula*) and *Pinus sylvestris*. Where both species are common, so is the disease. The blister rust of five-needled pines, *Cronartium ribicola* which alternates with species of currants (*Ribes*), has already been mentioned. Since it is virtually impossible to eliminate the alternate hosts, American five-needled pines are impossible to grow on any significant scale in Europe.

Apart from forest nurseries, where the use of pesticides is routine, and with the exception of *Heterobasidion annosum*, it is difficult to find good examples of successful protection against fungal epidemics in Europe, perhaps because they are so rare. In New Zealand the control of Dothistroma blight (*Scirrhia pini*) on *Pinus radiata* is achieved by aerial application of copper sulphate. This has made it possible to grow the species in a healthy condition, even when the local climate is favourable for the fungus.

By following the outbreak of fungal attacks, there is much to learn about the interaction between a tree and its diseases. An example is the attack by the canker *Gremmeniella abietina* on *Pinus contorta*, first planted in northern Sweden on a large scale during the 1970s (Karlman *et al.* 1994). Studies of *Gremmeniella* epidemics indicated clear correlations between severe weather conditions, site factors, unsuitable provenances, and the attacks. New rules governing the use of *Pinus contorta* were introduced, restricting its use to a maximum of 28 000 ha of new planting a year at low elevations in northern Sweden. Fine-textured soils must be avoided (Anon. 1994).

Preventative measures against insects

Forest hygiene, in the sense of not leaving too much material for beetles to breed in after harvesting or windthrow, is most important to prevent an outbreak of bark beetle (*Tomicus piniperda*) damage to healthy trees. Among other measures to minimize the impact of insect damage are:

(1) maintain populations of predators of potentially damaging insects, suggesting that excessive cleaning and removal of dead trees should be avoided;

(2) selection of species and provenances that will thrive on the site—unhealthy trees are much more likely to succumb to attacks;

(3) good site preparation also ensures healthy growth and reduces stress to the trees.

Chemical treatment of seedlings against some insects is quite common. Among these is the large pine weevil (*Hylobius abietis*, Fig. 11.4) which breeds in the stumps and roots of recently felled conifers. Coniferous seedlings have the root collars treated with an insecticide, usually before replanting, to prevent girdling by the weevils. Total loss of the crop can occur if protective measures are neglected. Successful protection during the vulnerable first 1.5 to 3 years is important—the period of development of a weevil from egg to adult. Some insecticides fulfil this function, but those currently permitted in Scandinavia are not sufficiently persistent and foresters are encouraged to use non-chemical methods of control.

As mentioned on page 69, scarification of a site can reduce weevil damage in regions where the use of chemicals is unacceptable, as can a delay of 2 to 3 years before replanting, though this is seldom considered a viable option.

Fig. 11.4 The pine weevil (*Hylobius abietis*) causes more damage to replanted conifers, by girdling young trees, than all other insects together in northern Europe. (Photo: R. Axelsson.)

Control

Frequently, damage is not serious enough to warrant any particular control measures or alternatively control can, in some cases, be so costly that it cannot be justified. This is true in the case of the green spruce aphid, *Elatobium abietinum*, an *r*-pest, in its periodic non-fatal attacks on *Picea sitchensis* in Britain (Hibberd 1991). In such cases, reduction in yield of a crop is inevitable.

Chemical control

Spraying insecticides or fungicides onto insect or fungal outbreaks in mature forests is exceedingly rare today. Only in the most exceptional circumstances are insecticides applied to reduce populations of harmful insects, commonly leaf eating insects. The larvae of the nun moth (*Lymantria monacha*) have caused sufficiently serious damage to require spraying. The long-term effectiveness of this treatment is questionable because it is not selective, killing the predators of the insect at the same time.

Biological control

Among the various possibilities for biological control of some moth and sawfly larvae, the use of nuclear polyhedrosis viruses (NPV) is most attractive, offering selective control of epidemics. Although many potentially damaging insects could be controlled in this way, only six are registered world-wide (Speight and Wainhouse 1989). Among these, aerial applications of emulsions of a parasitic virus have proved successful for 30 years in controlling the European pine sawfly (*Neodiprion sertifer*).

Severe outbreaks of bark beetles can be controlled using traps baited with appropriate synthetic pheromones. These chemicals are very similar to the pheromones produced by the insects to attract the opposite sex. Reproduction rates can be vastly reduced by using pheromone traps. Another partly biological means of control against appropriate insects is to use 'trap trees' that are injected with a systemic insecticide in an attempt to prevent the spread of an epidemic to healthy neighbouring stands.

However, the most common approach to an outbreak of a harmful insect is to wait, observing and measuring the size of the population. With some intermediate pests the population of the predators may gradually increase to a size when it will control the pest. Only if it does not may action be needed. With *r*-selected pests, the population usually collapses eventually from diseases of its own. Once serious damage to a plantation is already evident it is often too late to do anything about it.

Environmental aspects of wildlife, pest, and pathogen control

Planted monocultures can be more at risk than natural forests (see Chapter 2) because, when conditions are suitable for an attack, there are huge excesses of food upon which pests can grow and reproduce rapidly. Speight (1983) states that in virtually all examples, increasing the complexity and diversity of a crop system promotes the survival and success of parasites and predators. Diversity may eventually come to be seen as the best first line of defence in reducing the abilities of pests to proliferate. Other pest control techniques should then be required less often and on a smaller scale.

Measures of diversity can be achieved in even-aged plantation forests by:

(1) creating mosaics of different species and ages, even though each may be an even-aged monoculture;

(2) conserving riparian zones;

(3) establishing appropriate edges to the forest;

(4) the introduction of multistoried stands, though this will rarely be practicable;

(5) where possible, using long rotations;

(6) encouraging naturally regenerated saplings of other species in pre-commercial or other thinning operations.

The use of insecticides and fungicides is becoming unacceptable in forestry today and is likely to become increasingly so during the twenty-first century. The future for controlling epidemics seems to lie in the use of very selective biological control agents that are being tested for some pests: viruses against sawflies, nematodes against bark beetles, and predatory insects against aphids.

Conclusion—Integrated pest management

The most usual method of control will be to adopt an integrated pest management (IPM) approach, described in detail by Speight and Wainhouse (1989). This involves not only applying control measures when outbreaks occur, but reducing the susceptibility of the forest through silvicultural means—selecting species correctly, using vigorous pest-free planting stock, removing susceptible individuals in thinning operations, and other such measures. Though this may all seem obvious, success depends upon the co-operation of all forest owners, and may require government support and legislation.

12   Protection from wind

Wind damage to forests is a recurrent hazard in many parts of the world. High winds can uproot trees, break stems, cause malformations, and retard growth. They can also disrupt planned management, causing economic losses from higher logging costs, damage during storage, unharvested wood, lowered increment in residual stands, and shortened rotations, besides reducing landscape quality and habitats for wildlife.

Forests in temperate maritime regions suffer most seriously from wind damage, but those in continental mountainous regions can be badly affected too. In some places wind may be the ecological factor most responsible for the species composition of natural forests through destruction by storms. This leads to successions which favour colonizing, rather than climax, species. Storm damage is a constant risk, rather than an unforeseeable event, in many places where it occurs. It is the most important influence affecting the viability of commercial forestry in the British uplands.

Damaging winds

No tree can escape damage from a violent storm, where wind speeds exceed 30 m s^{-1} for 10 min or more (Mayer 1989). Sustained mean wind speeds of 25 to 29 m s^{-1} can cause considerable damage to trees on almost any soil. Gusts over 22 to 27 m s^{-1} can uproot individual trees or small groups on soils where rooting is restricted. Such gusts occur even when the mean wind speed is about 18 m s^{-1}, the lower limit of a gale on the Beaufort scale.

The nature and effects of airflow over plant surfaces have been described by Leyton (1975), Grace (1977, 1983), and more generally by Quine *et al.*(1995) and Coutts and Grace (1995). Normally, wind speeds fluctuate irregularly between lulls and gusts; these fluctuations are caused by turbulence and it is this rather than mean wind speed which is mainly responsible for damage. Turbulence is an extremely complex phenomenon (Fig. 12.1), but generally the higher the wind speed and the rougher the forest canopy aerodynamically, the greater the degree of air turbulence. Usually wind has relatively little harmful effect on the canopy, especially when there is crown contact between neighbouring trees, but turbulence

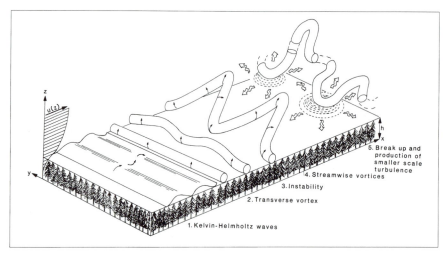

Fig. 12.1 Schematic representation of the formation of damaging gusts over forests. (From Finnigan and Brunet 1995.)

readily causes stems and branches to sway and also root movements in the soil. If lateral roots are not secure, some movement occurs in the root plate which then rises and falls. Eventually the trees may be uprooted. The random nature of turbulence is often illustrated by the haphazard directions in which tree stems are thrown by the wind, irrespective of its direction.

Crops at risk

Quine (1992) recognized three components of vulnerability: persistent, progressive, and episodic. Persistent vulnerability is determined by factors that remain relatively unchanged throughout the life of the stand, such as soil or method of cultivation. A progressive change in vulnerability with age occurs by changes in tree form, including height and height/diameter at breast height ratio. Episodic changes in vulnerability can be brought about by silvicultural operations—thinning, respacing, seeding fellings—as well as other events such as snow-loading, high rainfall, and whether the trees are leafless or not at the time of the wind. Silvicultural operations carried out late in the rotation on sites with high, persistent vulnerability can substantially enhance the risk of damage. Even very young crops can suffer from wind damage, especially if they are planted in individual plastic shelters which are blown over.

Winds are obviously the direct agent of damage so the crops most at risk are clearly located in the windier regions of a country, especially at high elevations where there is little topographic shelter and where soils provide poor rooting conditions. Large-scale topographic features can funnel winds to produce recurrent damage, and smaller-scale topographic features determine the precise points affected by windthrow (Jane 1986). In continental regions, increased damage, due to increased wind speed and turbulence, is often localized. It is linked with summer thunderstorms and with föhn winds close to mountain areas, caused by temperature inversions. The highest winds occur near isolated mountains and especially where airflow is funnelled through valleys. High mountains can augment the force of winds towards the bottom of slopes, where serious squalls may occur. Quine and Miller (1990) conclude that föhn winds explain the concern with lee-slope damage in much continental European literature (e.g. Hütte 1968). Lee slopes can provide shelter to susceptible sites during normal weather conditions, but susceptibility can be increased during storms (Jane 1986).

Crops on soils which permit only shallow rooting, especially wet soils and those which have a low sheer strengths, are particularly vulnerable, as are recently thinned stands which are unable to dissipate the energy of the wind by crown contact.

The onset of damage appears to be earlier in conifers and increases more rapidly with age than in broadleaves. *Pinus*, *Picea*, and *Abies* are specially at risk. This must be attributed, in part, to the establishment of these genera on exposed sites as much as to the inherent characteristics of the trees. Broadleaved species are usually leafless during stormy periods of the year and it could be argued that they are therefore much less at risk. However, Quine (1989) considers that it is by no means certain that the leafiness of trees will, in all cases, exacerbate wind damage. An intact canopy results in reduced penetration of the wind, at least at lower wind speeds, which could reduce the forces acting on stems for a given wind speed experienced above the canopy.

The precise elements of crop structure which influence energy transfer are not clearly understood and are the subject of continuing research. Obviously crown size, density, and tree height are important, but so too is the spatial distribution of trees and topography (Miller *et al*. 1987; Ford 1980). Mayer (1989) states that conifers can absorb energy from turbulent winds only within certain frequency ranges if they are to escape damage. Vulnerability always increases with height, and therefore age. In extreme conditions, stands of 5 m or more may be at risk, though 10 to 15 m is more commonly the threshold value. The turning moment exerted by wind of a given velocity on a crown is certainly much greater in a tall tree than in a short one. Trees with well-tapered crowns which occupy a large proportion of their total height have lower bending moments than short-crowned trees of the same height, and so are more stable.

It is known that trees are more easily uprooted or broken if they are very slender; that is, if the taper in the stem decreases only slightly with height. This is because the ratio of the bending moment produced by the wind to the resistive moment produced by the elasticity of the stem increases if taper is slight (Petty and Worrell 1981). Taper is more pronounced in dominant trees and in thinned crops, and is less in unthinned crops once competition has started and in less dominant trees

(Newnham 1965). Hence crown breakage is more common in codominant trees than in dominant trees. Slenderness is one of the factors which is most easily manipulated by foresters, by spacing and thinning correctly. Wood density may also have an effect on susceptibility to windthrow and windbreak. Nepveu *et al.* (1985) found that after a storm, the density of 66 per cent of the broken stems of *Picea excelsa* was lower than the standing controls, especially among smaller diameter trees.

Anything which injures tree roots or infects them can predispose them to risk. This includes soil compaction by machinery during the usual forest operations, root rots, and waterlogging.

Damaging effects of winds

The most spectacular and catastrophic form of wind damage occurs as a result of major storms and tropical cyclones which, fortunately, are infrequent—perhaps 50 to 100 years apart. They usually devastate quite small areas which lie in the tracks of the strongest winds. Nothing can be done to prevent losses, for if trees are not uprooted, stems or branches are snapped, usually near the bottom, the most valuable part of the tree (Fig. 12.2). Relatively recent examples of catastrophic damage were the now well-documented storms of November 1972 and April 1973 which damaged some 7000 ha of broadleaved trees and 65 000 ha of conifers in Lower Saxony and neighbouring areas of Germany (Kleinschmit and Otto 1974), The Netherlands, and Denmark. Wind speeds of up to 48 m s^{-1} (Force 16) were recorded. About 19 million m^3 of timber were damaged in the former West Germany of which 15 million m^3 were in Lower Saxony and amounted to almost 12 times the sustained annual cut of the region (Anon. 1973). The storm of 16 October 1987 was one of the worst to affect southern Britain (where 15 million trees, or 4 million cubic metres of timber were uprooted) and neighbouring parts of the continent when maximum mean gusts of 41 m s^{-1} were recorded (Quine *et al.* 1995).

In Britain, the return period for major storms in any one place has been estimated by Shellard (1976) as about 50 years. Longer or shorter intervals have been calculated for other regions. For example Lorimer (1977) estimated a recurrence interval for large-scale windthrows as 1150 years in north-east Maine, USA, this being much longer than the interval needed to attain a climax, all-age structure. The period may only be about 10 years in parts of the South Island of New Zealand (Hill 1979).

At the other extreme of the spectrum of damaging winds are those likely to occur at least once each year. These are less dramatic, but often more serious in their effects on trees in terms of the insidious attritional damage they inflict, in which individual trees or small groups are uprooted. Once this type of endemic windthrow begins, it may extend quite rapidly (Fig. 12.3) and often results in the need for premature clearance of whole crops, long before the desired rotation is reached. In Britain, about 15 per cent of annual production comes from windthrown timber.

Fig. 12.2 Serious storm damage to spruce, causing widespread windsnap. (Crown Copyright.)

Endemic windthrow often occurs during the period in the life of a crop when current annual increment is at its highest. If such windthrow can be delayed for only a few years, total production can be significantly increased as well as the proportion of larger-sized logs (Fig. 12.4).

In less extreme conditions tree crowns can be damaged by windbreak of tender leading shoots which results in forking and other malformations. Certain species, coastal provenances of *Pinus contorta* and some other pines are notorious examples, do not root well when very young and become prone to toppling, which causes undesirable curves in the base of the stem and the development of reaction wood. Later they become even more susceptible to both windthrow and snow damage (Petty and Worrell 1981).

Wind can influence the rate of growth of trees. In controlled environment experiments, Rees and Grace (1980*a, b*) found that height growth of *Pinus contorta* was reduced by 20 per cent by wind speeds of up to 8.5 m s^{-1}, though radial growth was not affected. This was ascribed to the shaking effect of wind rather than water stress and was accounted for by differences in cell length rather than cell number.

Fig. 12.3 The start of windthrow in Sitka spruce (*Picea sitchensis*) on a surface water gley. Trees tend to be uprooted along the lines of the plough ridges. Note the very shallow rooting.

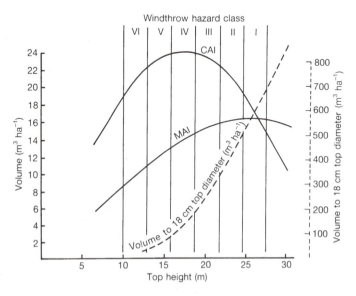

Fig. 12.4 Relationships between growth of yield class 16 m³ ha⁻¹ year⁻¹ Sitka spruce (*Picea sitchensis*) and top height (data from Edwards and Christie 1981). The stage of onset of windthrow in various hazard classes is also shown (data from Miller 1985).

Silvicultural methods of reducing windthrow risk

There are three approaches to combating endemic windthrow (Ford 1980):

(1) improving root development;

(2) designing a form of silviculture appropriate to short rotations;

(3) improving forest design.

In practice more than one of these approaches is usually taken by those who manage windthrow-prone forests, and they are reviewed by Savill (1983). A further option is to abandon forestry on the worst sites—a choice always worth considering where there is an alternative form of land use, such as sheep grazing.

Improving root development

Ground preparation

An obvious approach to delaying the onset of windthrow is to improve the anchorage of trees by increasing the rooting depth and the radial spread of roots, even by small amounts. This is discussed in detail by Quine *et al.* (1995).

The root form of some species may be determined by the soil type in which it grows. By far the most difficult soils are wet ones such as silty and clayey gleys whose textures impose permanent limitations on rooting depths due to lack of oxygen. On these soils the great majority of tree roots, even in drained areas, are shallow and severely restricted at about 10 cm below the soil surface. The possibilities for stabilizing crops on such soils are relatively few. Attempts to modify one contributory cause of windthrow, that of waterlogging, through intensive drainage by spaced furrow ploughing has introduced another problem of restricting root spread, which also contributes to stability. Among the most promising techniques for combining reasonable soil drainage with wide root spread and marginally better root depth is the use of mole drainage on gleyed soils and tunnel ploughing on peats (p. 73), methods which are being widely employed in new afforestation schemes in Ireland. On wet replanted sites, Quine *et al.* (1991) have shown that a former practice of planting on the slightly elevated soil very close to old stumps is likely to result in instability later in life. This is because the presence of a stump closer than 70 cm to the planting position can prevent the development of a uniform root system and cause clustering of roots away from the stump. Mounding is now commonly practised to overcome this problem (see Chapter 5).

On drier but compacted soils, and those with ironpans, there is often a very sharp divide between the upper soil in which the trees root and the compacted soil below. A plane of weakness develops which can make crops very prone to windthrow. Deep ripping can aid root development and result in much improved stability on these sites.

In the long-term, breeding trees for improved root ventilation which would enable them to root more deeply in waterlogged soils (Sanderson and Armstrong 1978) or other characteristics, such as root-regenerating capacity (Nambiar 1981) or possibly a flatter cross section, which is said by Brinar (1972) to confer improved stability, could become important.

Planting stock

Improvements to the rooting qualities of planting stock and methods of planting are an important method of making some species more stable. Some *Pinus* spp., *Pseudotsuga menziesii,* and *Larix* lose their capacity to initiate first-order lateral roots, which eventually become the main structural elements, early in their first season of growth (Burdett 1978, 1979). Thus, the final configuration of the root system may be established very early in life and remain unchanged with increasing age. Bad planting can lead to instability. One possibility for overcoming these problems is to use small container-grown plants (Chapter 7) which can be of special value in circumstances where root systems can be developed which grow in a similar way to the roots of naturally seeded trees.

Species selection

There are obvious theoretical advantages in growing species which are likely to be windfirm. Trees which root deeply, and have buttressed trunks and wood which is strong and resilient are most likely to have these characteristics, but unfortunately they are possessed by few of the commonly-used plantation species. The drag coefficients of broadleaved deciduous trees can be reduced by a factor of 10 when they are leafless (Quine *et al.* 1995) which tends to be during the period of the year when storms are common. Broadleaved species therefore have obvious attractions, especially as they frequently root more deeply in wet soils (Gill 1970). Unfortunately, at least in plantation forestry, they are less commercially attractive because of their lower levels of production on poorer soils. Some broadleaves, such as *Juglans* and *Paulownia* have reputations for being susceptible to wind damage as do some *Populus* clones such as *P. trichocarpa* 'Fritzi Pauley'.

Some species can escape damage even from cyclones because of their ability to shed leaves and twigs before wind forces are sufficient to cause windthrow or branch or bole fracture. Several tropical *Eucalyptus* spp. and species of *Araucaria* have this reputation (Stocker 1976; Brouard 1967). It is an unfortunate fact that some of the commercially most important species in much of Europe are *Picea* spp., *Abies* spp., and *Pinus* spp. which present large, stiff, and resistant foliage areas to the wind. They have much higher drag coefficients than, for example, *Pseudotsuga menziesii*, *Larix* and *Tsuga heterophylla*, with their flexible branches which stream out in the wind (Raymer 1962; Walshe and Fraser 1963). The latter three species are widely planted, or are gaining in importance in parts of continental Europe. However, drag coefficients vary both within species and with wind speed (Mayhead 1973). In many reports of windthrow, *Picea* and *Pinus* are reported as being the most damaged species. However, significant differences in susceptibility to wind damage sometimes occur between provenances. In *Pinus caribaea*, provenances from coastal origins in central America, which experience frequent cyclones, are more resistant than inland provenances (Forestry Department Queensland 1981).

Root and stem rots

Trees infected with various butt and root rots usually have many dead or dying roots and weakened stems which make them more susceptible to windthrow and windbreak than sound trees. In the case of *Heterobasidion annosum* this was established as early as 1937 by Bornebusch, in Denmark. First-rotation sites are often free of such diseases. Though freedom is unlikely to last indefinitely, the use of relatively resistant species and at least delaying the onset of infection by careful stump treatment is important.

Silviculture appropriate to short rotations

Silvicultural systems

Quine and Miller (1990) have developed a set of recommendations for silvicultural systems in relation to windthrow vulnerability. They considered that where risks are greatest, the most widely applicable, suitable system will be a clear-felling without any commercial thinning. Ideally the plantations should cover a wide range of age classes, and careful attention should be given to the management of edges. Selection and coppice systems could be used too, but are unlikely to be widely applicable. All other silvicultural systems are likely to be unsuitable, especially uniform shelterwood and group systems.

Rotation length

The probability of wind damage can be reduced if rotation lengths are sufficiently short that crops spend little, if any period, at heights which expose them to risk. Rotations aimed at achieving sawlog-sized material by the time the trees reach 20 m top height have been suggested for parts of Britain through the use of heavy fertilizing, wide spacing, and early precommercial thinning.

Spacing

Two mutually incompatible aspects of stability related to spacing and thinning have already been mentioned. On the one hand, damaging wind turbulence is reduced by having a relatively smooth canopy to the forest, implying narrow spacings; on the other, the maintenance of well-tapered stems which bend with the wind rather than break or become uprooted requires wide spacings.

A great deal of continental European and Australian literature dwells on the importance of maintaining well-tapered trees, slenderness usually being expressed as a height/diameter at breast height ratio (e.g. Braastad 1978; Brünig 1973; Sheehan *et al.* 1982). Many consider the maintenance of an acceptable ratio to be by far the most important influence on stability. Faber and Sissingh (1975) consider that in The Netherlands a height/diameter ratio of no more than 50/1 or 60/1 is necessary for adequate wind resistance. Kramer and Bjerg (1978) suggest that a ratio of about 80/1 might be acceptable. To maintain a low height/diameter ratio, either very wide spacings with no subsequent thinning or early and heavy thinnings are necessary: if height/diameter ratios are allowed to reach high levels, late thinning is unlikely to reduce them significantly.

Traditional treatments involving thinning, which are enshrined in many yield tables, are criticized by Brünig (1973) as leading to dangerously high height/diameter ratios. The price of greater stability, implying a lower stocking than is customary (Table 12.1), is that a lower yield has to be accepted. An alternative view is that close spacing and the maintenance of a relatively smooth canopy are more important for stability. Richter (1975) in a study on *Picea abies* in Sauerland, north-west Germany, found a tendency for damage to decrease with increasing stand density as did Etverk (1971) in Estonia who concluded that dense stands of relatively slender trees are more stable than more widely spaced stands. Sutton (1970) in New Zealand, Bornebusch (1937) in Denmark, and many British foresters (e.g. Fraser 1964) advocated rather dense crops especially where conditions are not good for rooting. It is possible that when snow and ice are present in combination with wind, wide spacings are safer, but where the problem is that of wind alone, especially on very wet soils, narrow spacings may be less at risk.

Thinning

The dangers associated with thinning are well understood. Thinned crops are more susceptible to damage during the 1- to 5-year period when the canopy is reclosing,

Table 12.1 Height/diameter ratios at 20 m top height in unthinned Sitka spruce (*Picea sitchensis*) (based on Kilpatrick *et al.* 1981)

Stems (no. ha^{-1})	Top height/dbh* ratio
500	64
1000	72
1500	80
2000	88
2500	96
3000	104

*dbh = diameter at breast height.

when they are unable to dissipate the turbulent energy of the wind by crown contact. Cremer *et al.* (1977), for example, found that with *Pinus radiata* near Canberra, stands over 30 m tall which had not been disturbed by recent thinning, had only 0.1 per cent of the trees thrown in a gale in July 1974 whereas 11 per cent of the trees in stands thinned in the previous 5 years were damaged.

Forces on a tree can be doubled by removing its neighbours (Walshe and Fraser 1963). Experience with frequent gales in upland Britain has shown that thinned crops have a decreased height expectancy of about 3 m over unthinned. Furthermore, once 3 to 4 per cent of a stand is windthrown possibly 50 per cent of the remainder will succumb within the next 3 m of height growth (Mayhead *et al.* 1975). There are therefore obvious advantages in not thinning at all or giving up thinning very early. Such regimes are adopted in about half of the forests owned by the British Forestry Commission (Ford 1980) and three-quarters of these in Northern Ireland (Savill and McEwen 1978).

On high-risk sites that are thinned, regimes that cause large openings in the canopy, such as heavy or systematic thinnings, result in more damage than light or selective thinnings. Delayed thinning and especially heavy thinning in older crops which involves releasing tall slender trees with high height/diameter ratios is often quoted as being an important contributory cause of damage. The removal of such trees in early thinning to increase resistance to both wind and snow damage is advocated by some authors (e.g. Slodicak 1987).

Where conventional thinning is very risky, much attention is paid to maintaining timber quality by adjusting initially rather dense spacings through precommercial thinning—that is thinning when crops are no more than 5 or 6 m tall, well before they are exposed to any serious risk. The use of thinning techniques that never allow canopies to be opened sufficiently to endanger the crop are also receiving attention. These include the use of self-thinning mixtures of slow- and fast-growing species, or of slow- and fast-growing provenances of the same species (Lines 1996), and using arboricides such as glyphosate to inject into unwanted trees (Ogilvie and Taylor 1984).

Pruning

The effects of high pruning on stability, especially the pruning of edge trees, are not clearly known and require more investigation. Some workers believe that it helps by allowing the wind to filter through the crop and so reducing turbulence, as well as reducing the sail effect of large crowns by reducing swaying.

Forest design

Classifying risk

The extent of wind damage can be influenced by plantation layout and local topography. It is, however, extremely difficult to take these factors into sufficient consideration when establishing large areas of forest.

Several authors have produced hazard classifications. J.F. Miller (1985), revised by Quine and White (1993), for example, recognized six hazard classes for Britain based on four site variables: wind zone determined from a map, altitude, exposure, and, most importantly, soil. The system is used for zoning coniferous forests, especially *Picea sitchensis*, of 500 ha or more, and from it top heights at which windthrow may be expected to start are indicated for each hazard class. Adjustments to the system allow for the effects of thinning and method of ploughing to be taken into account. The classification provides a basis for delineating broad no-thinning or early thinning areas.

Experience elsewhere may allow much more detailed action. For example from a study in the German Harz mountains, Hengst and Schulze (1976) proposed measures to increase wind resistance, including the creation of new road networks with reconstructed stand margins, the conversion of large plantations to a varied wind-resistant mosaic of different species and ages, and the progression of fellings against the prevailing wind with an elastic time sequence of felling.

Exposed edges

The edges of forests are a particular cause of vulnerability. For reasons discussed below, it is certain that unsuitable edges will predispose crops to damage but the reverse is not necessarily true. There have been many attempts in the past to create windfirm edges in the hope that they will protect the crop behind. Most have been notably unsuccessful—the edges often remain stable, but the crop behind is as vulnerable as ever, or more so.

The danger from edges arises because relatively large wind eddies and much greater bending moments are produced within a distance of 10 to 15 times the height of the edge. There is a more extended region beyond, up to 40 to 50 times the height, in which wind turbulence gradually declines. Edge trees start the enhancement of turbulence but themselves remain largely intact (Hill 1979). In fact, very high mean wind speeds of about 36 m s^{-1} are required before any damage occurs to trees on permanent edges (Papesch 1974).

The treatment of stand margins has therefore received much attention. Kramer (1980), for example, stated that a depth of 30 to 50 m from edges should be specially treated as a shelterbelt. It should be heavily thinned, beginning at 8 to 10 m, and not thinned after 15 m top height. This will encourage potentially stable, well-tapered trees with long crowns. He also emphasized the need for wide spacing in edges, all with the aim of obtaining an edge zone which wind can penetrate, instead of a dense impenetrable border. Fraser (1964) has shown that these treatments have the effect of eliminating the zone of fluctuating forces behind the edge. Mitscherlich (1973) and Gardiner and Stacey (1996) similarly emphasized the adverse effect of dense margins and proposed wide spacing and pruning to allow the wind to filter through the edge zone. Hütte (1968) and Neckelmann (1981, 1982) tried cutting the tops off trees within 15 to 25 m of the edge zone, with lorry-mounted hydraulic shears, to produce a slowly rising zone. While this reduced damage to the edges themselves, the plantation behind still suffered from wind-throw, but the topped edge trees often died within a year or two of cutting the crowns and their value was totally lost. Recently-cut edges, on the boundaries of clear fellings, have none of these advantages and are frequently severely damaged. Little can be done about them, though experience in Britain has been that if the temptation to tidy them can be resisted, damage does not always extend too quickly. Internal breaks in the forest along which roads are constructed should be 30 to 40 m wide to encourage low height/diameter ratios of the edge trees.

Species mixtures

Ford (1982) has stated that the classic response to catastrophe is to diversify the forest by planting species which are less vulnerable. This reaction was very clear in Germany following the 1972/73 gales where areas of *Picea* and *Pinus* have been markedly reduced in favour of replacement by broadleaved species.

Intimate mixtures or occasional belts of deep-rooting deciduous trees in crops of shallow-rooting evergreen conifers are also frequently used in an attempt to promote stability, with varied and often indifferent success. Experience in Britain has often been that the deciduous component of *Alnus* or *Betula* remains stable and the conifer, usually *Picea*, is at least as vulnerable as before. It is possibly more vulnerable because the deciduous trees are usually leafless at the times of major storms and cause pockets in which wind turbulence can increase.

Size of clear-felling coupes

For equivalent areas of standing and felled trees, damage is consistently less adja-cent to large clear felled areas than to small ones. This is largely explained by the fact that per unit area felled, a few large coupes will have a smaller perimeter length of edge trees at risk than several smaller ones (Gordon 1973; Neustein 1964). For example a square felling coupe of 10 ha will have a perimeter length of almost 1.3 km, whereas five square coupes of 2 ha each have perimeters totalling 2.8 km, or over twice the length of the single 10-ha coupe.

When clear felling it is usually safest to fell right up to established edges along roads, water courses, young stands, etc. or to open ground, and as a poor second best, to proceed with felling against the direction of the prevailing wind.

Age diversification

To avoid local catastrophes from windthrow, there is a strong case for deliberately trying to spread the risks by diversifying the age classes of stands in the forest. This action substantially reduces the proportion of trees at risk at any one time. It can also minimize the pressures on local harvesting and sawmilling and labour resources, which can be stretched beyond their capacities after large windthrows.

Webber and Gibbs (1996), for example, describe how 70 000 m³ of pine logs had to be stored on a continuously irrigated site, to prevent bluestain and decay fungi developing, for about 3 years after a gale in 1987 that blew down 4 million m³ of timber in the south of England (see Fig. 12.5).

Fig. 12.5 Storage of windthrown timber under water sprinklers, to prevent decay, at Thetford Forest, England. (Crown Copyright.)

13 Fire

Fire can be a spectacular and extremely destructive cause of damage to plantations. But, although the great majority of such fires in Europe are caused by human activities, fire has long been part of natural ecosystems, a point apt to be forgotten, though not irrelevant, when considering its threat to plantation forestry. A comprehensive account of the nature of forest fires in North America, their significance and their management, including fire fighting, is given by Pyne (1984). A great amount of work into the nature of fires and fire prevention in southern Europe, where fires occur every year, sometimes on vast areas, is currently in progress and present knowledge has been described in detail by Trabaud (1989), Delabraze (1990a), and Moreno and Oechel (1994).

Ecology of fire

In the boreal forests most natural stands that are cut are of post-fire origin. Indeed, much of their diversity in composition, vigour, and character is due to irregular but periodic fires. Fire acts as a primary nutrient cycling and rejuvenating mechanism, by releasing nutrients bound in organic matter where otherwise mineralization is slow due to low temperatures, drought, or acidity, and stimulating the growth of nitrogen-fixing plants. The importance of fire in forest ecosystems is discussed in detail by Kozlowski and Ahlgren (1974), Wein and MacLean (1983), and Kilgore (1987). The idea that the artificial exclusion or suppression of fires should not occur in natural areas was widespread in the United States until the catastrophic wildfire which started in Yellowstone in 1988. It burnt 32 000 ha of the park and 170 000 ha of forests outside it and threatened people and settlements (Calabri and Ciesna 1992). Since then a great amount of research has been devoted to managing natural ecosystems in harmony with controlled natural fires. The systematic suppression of fires has been shown to have several detrimental effects—the main one being the build-up of large amounts of fuel which can lead to wildfires getting out of control.

Trees react in different ways to fire and some subclimax forests are effectively propagated by it. Rowe (1983) classified trees according to their reactions as invaders, evaders, resisters, endurers, and avoiders of fire.

Invaders are typically *r*-selected species which produce enormous quantities of wind-disseminated seed. They colonize and grow rapidly on recently burnt sites and include such genera as *Alnus*, *Populus*, *Salix*, *Betula*, and *Pinus halepensis* (Trabaud 1981).

Evaders of fire store quantities of strongly dormant seed, either in serotinous cones in the canopy or in the soil. Germination is triggered by the high temperatures of fires. Thus, in the northern part of their ranges, in Canada and the United States where fires are common, the cones of Jack pine (*Pinus banksiana*), lodgepole pine (*Pinus contorta*), and, to a lesser extent, black spruce (*Picea mariana*) will not open unless the resins which hold the scales shut are subjected to great heat. In Europe, *Pinus halepensis* and *P. brutia* also belong to this category (Naveh 1975; Saracino and Leone 1994).

Resisters of fire either have very thick bark, for example *Pinus pinaster*, *P. pinea*, and *Quercus suber*, as protection from ground fires, or survive through the presence of dormant buds in the stem, for example most *Quercus* spp. and some *Eucalyptus* spp., enabling renewal of the crowns of scorched trees.

Similar mechanisms are found in fire endurers. Even if totally burnt above ground, several *Populus* spp. and *Alnus* spp. regenerate from root suckers, *Quercus ilex* from sprouts, and a few *Eucalyptus* from lignotubers.

Only species that normally avoid fire can be eliminated by it. These are mostly late successional species, normally strongly shade-bearing, such as many *Picea* spp. and *Abies* spp., which occupy unburnt areas and thrive where fires are very rare or absent.

These adaptations to fire can, from time to time, be both a help and a hindrance when coping with fire damage in plantations. If a young hardwood crop such as *Quercus* or *Castanea sativa* has been burnt, often the vigorous regrowth from the base can be used to advantage for restocking the ground. By contrast where a crown fire has burnt through a pole-stage *Pinus* plantation the dense natural regeneration that sometimes springs up can create problems of what to do with it—to clean, to respace, or to clear and replant?

Losses from fires

Average annual losses from fires are close to 0.5 Mha in Europe (Calabri and Ciesla 1992), though there are very large between year variations: in France 6700 ha were burnt in 1988 but over 75 000 ha in 1989. However, they mostly occur in natural ecosystems in which the tree cover is sometimes sparse. For instance French statistics include both forest formations, where tree cover is over 10 per cent and 'subforest' vegetation which includes heathland and Mediterranean 'maquis' and 'garrigue', and in Italy all vegetation fires are counted as forest fires. Over 80 per cent of the area burnt in France in an average year is 'subforest' vegetation, but in bad years forests can make up to 50 per cent of the burnt area (van

Effenterre 1990). Although in some remote areas, mainly in North America and in Australia, bush fires are left to burn with no human intervention; in Europe it is generally preferable to control fires because of the high population density and the proximity of human habitation or industry.

Burning of plantations is small by comparison but, of course, industrial plantations and intensively managed forest are of high economic value and represent a considerable investment. In the two worst years in the decade 1981–1990 in France—75 000 ha burnt in 1989 and 72 000 ha in 1990—two fires burnt respectively 3600 ha and 5600 ha in the highly productive Landes massif, which had the reputation of being well managed for fire prevention.

Most European forest fires occur in the south (Table 13.1); 80 per cent of French forest fires occur in the Mediterranean region. However, even Brittany is affected with fires, with, on average, more than 0.5 per cent of the forested area burnt annually.

Apart from losses of wood, destruction of forests by fire can lead to seriously increased water runoff with consequent dangers of flooding, erosion, and siltation. The recreational and amenity values of forests are reduced and fires can endanger agricultural crops and buildings. Contrary to popular belief, most vertebrate wildlife escapes fires by flying, running, or burrowing underground. The main effect of the fire is on the habitat which, after a short period of recovery, is usually enriched in both numbers and diversity of species (Wright and Bailey 1982; Fox 1983).

Susceptibility of forests to fire

It is useful to consider the susceptibility of forests to fire danger in two ways:

(1) the fire hazard, the condition of the forest and how easily the vegetation and trees can be burnt;

(2) fire origin, which causes a fire to start.

Table 13.1 Average yearly burnt area in the 1981–1990 decade in Europe (data from Calabri and Ciesla 1992)

Country	burnt area (× 1000 ha)
France	49
Greece	38
Italy	61
Portugal	81
Spain	191
Other (southern Europe)	18
Other (northern Europe)	10
Total Europe	448

Fire hazard

Climate

Conditions of drought, low atmospheric humidity, high soil surface temperature, and, most of all, high wind increase the fire hazard.

Throughout northern Europe dry polar continental air in spring, accompanied by strong easterly winds and low precipitation, makes the period from January to April (February to May further north) the one of greatest danger because there is much dry grass (Fig. 13.1) from the previous season before new growth begins (Parsons and Evans 1977). A second, less serious peak in fire danger occurs in high summer as a result of high temperatures and low humidity which makes the undergrowth dry and flammable.

By contrast, in southern Europe some fires occur between January and March, due to poorly controlled forest cleaning operations or agricultural and pastoral activities (scrub burning), but the period of extreme fire danger is during July and August (Fig. 13.2).

Condition of the forest

The composition of a forest stand largely determines its susceptibility to fire. This is most easily understood from the idea of fuel loading. A fire will only burn if there is a sufficient quantity of material and if such material is reasonably

Fig. 13.1 Fire spreading from the grass *Molinia caerulea* to Sitka spruce (*Picea sitchensis*). Dead dry grass can easily be set alight during dry spells in early spring. (Crown Copyright.)

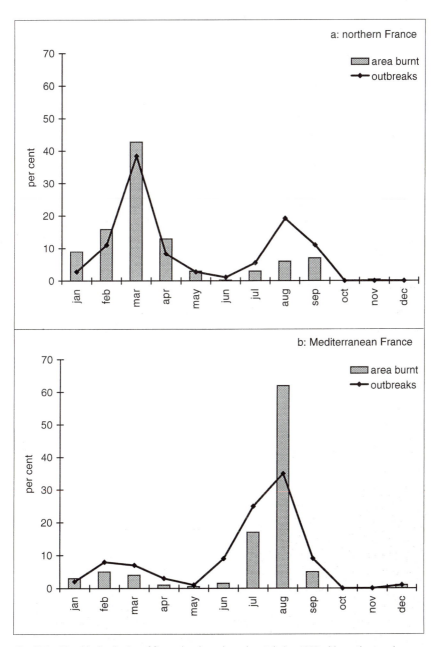

Fig. 13.2 Monthly distribution of fire outbreaks and area burnt during 1993. a) in northern and Atlantic France; b) in Mediterranean France. During this relatively favourable year, 1802 fires burned 4950 ha in Atlantic France, mainly in the Landes massif, and 2963 outbreaks burned 11 745 ha in the Mediterranean region. (From Stéphan 1994.)

flammable—lush green vegetation presents a very low hazard. Within a forest, flammable material comes from three sources: the trees themselves, the undergrowth, and the litter which accumulates on the forest floor; a fourth component can sometimes be added if the plantation is growing on peat, in which case the soil itself can burn.

Inside a forest stand the vegetation that develops under the trees is the greatest source of danger. Young trees growing in dense grass experience one of the most hazardous periods in the life of a stand. In northern Europe, most damage is done to such crops. Grass fires are sufficiently hot to kill the trees; they can also move with great rapidity. Once the canopy closes most herbaceous weeds are suppressed but under some species a considerable amount of undergrowth can continue to survive. In dry regions, shrubby species are often highly flammable (e.g. *Erica*, *Calluna*, *Phyllirea*). They represent the most serious hazard and make the crop potentially more flammable than one more densely stocked with trees but without the undergrowth—an important role of forest fire prevention is to eliminate this material.

Of the forest trees themselves, some species are more flammable than others. Valette (1990) has studied many tree and shrub species of southern Europe, and demonstrated the high inflammability of *Acacia* spp., *Alnus subcordata*, *Castanea sativa*, *Eucalyptus* spp., *Pinus halepensis*, *Quercus ilex,* and *Q. pubescens*. Conifers have the unjust reputation of being highly flammable; this derives mainly from the fact that the undergrowth of conifer-dominated forest is often composed of very flammable species. The less flammable species which are safer to plant in southern Europe to minimize the risk of a serious fire are *Cupressus arizonica* or *C. sempervirens*, or, on favourable sites, *Cedrus atlantica* and *Abies* spp.

The litter on a forest floor is an accumulation of dead foliage and branches. In many parts of the temperate zone litter breakdown is slow and up to 10 t ha^{-1} can accumulate. When dry and packed down it is a serious hazard. In general, coniferous litter decomposes much more slowly than litter from broadleaved trees (see Chapter 9).

Fire origin

Fire origin, or risk, is the other component of fire danger—that is, how likely is a fire to start and what are the causes? Although the causes of fires have been investigated in detail for the past 20 years in Europe, they are unknown for 55 per cent of outbreaks in northern France and 70 per cent in the Mediterranean area (Stéphan 1994). Statistics are often difficult to obtain and their interpretation is subject to caution, but co-operation between Mediterranean countries has led to the creation and use of consistent databases (Chevrou *et al.* 1995). Figures given here mainly concern temperate areas in France, which have been studied in detail (Stéphan 1994).

In nature, practically all fires are caused by lightning or volcanic activity. In managed plantations these causes are rare compared with fires started accidentally or deliberately by people. Only 2 per cent of forest fires in Europe are ascribed to lightning, though the proportion is somewhat higher in some sparsely populated northern countries—Finland 25 per cent, Sweden 12 per cent.

In northern France, 80 per cent of fires of known origin (representing 55 per cent of the area burnt) are caused by accident or carelessness. A great many of the accidental fires can be attributed to legitimate activities on neighbouring land. This includes agricultural practices—such as moor burning to bring on an early grass sward in the spring—which represent 47 per cent of accidental fire outbreaks (76 per cent of the area burnt by accident) in northern France. Forest operations also cause over 26 per cent of accidental fires. Other accidental causes include leisure activities (barbecues, cigarette butts) burning of rubbish heaps, sparks from road or rail vehicles, and sparks from electric transmission systems.

The most alarming fire risk is from arson. However these represent only 13 per cent of fires of known origin and 15 per cent of the area burnt in northern France. A detailed study in Mediterranean France showed that, contrary to a widespread belief which attributes the majority of fires to arson, fires from unknown causes are statistically very similar to those from known causes, indicating that statistics about fires of known origin can generally be applied to all fires—with some small corrections (Alexandrian and Gouiran 1990). Arson in the French Mediterranean area is no more common than in northern France, but in Corsica and in Italy it is much more widespread—50 per cent in southern Italy and 94 per cent in Sardinia.

These figures indicate that the main effort of prevention should be devoted to education of the public, and especially of agricultural and forestry workers.

Fire prediction

It will be clear from what has been said that when a high fire hazard and high risk occur together there is high fire danger. For example a young stand growing in dense undergrowth (high hazard) close to a picnic site (high risk) is highly susceptible. Climatic conditions then determine the overall risk—for example it will be very high during a strong wind after a prolonged drought.

Some very sophisticated ways of assessing the fire hazard have been developed which usually employ many climatic parameters such as soil water reserve or number of days since heavy rain, air and soil temperature, relative humidities, and wind speeds (Sol 1990). When predictions suggest highly desiccating conditions and strong wind in the spring in northern Europe or in summer in southern Europe, and when there is much flammable vegetation, fire danger is extreme. In northern Europe some periods—such as weekends or public holidays—present greater risks, as many people visit forests then. In the Mediterranean region, some areas have to be closed to the public during the summer period.

When extreme danger is predicted it is usual to broadcast warnings, activate procedures to reduce access to woodland, and generally publicize the great danger. All these are aimed at lessening the chance of a fire starting by reducing the risks of accidental causes or carelessness—which are the most important.

Nature of fires

Three stages of development can be identified for a damaging fire (Rothermel 1982).

1. Ignition and survival—most fires die quickly after ignition either because the ground and fuel are too damp or there is too little fuel to burn—the object of firebreaks.

2. Spread through surface fuel—a fire will develop where there is a plentiful supply of fuel such as dry grass, shrubs, litter, and brushwood. In porous fuels such as grass the spread can be very rapid. In compact fuels such as fallen logs it is slow.

3. Crown fires—in a crown fire the whole tree as well as the undergrowth is alight. In a dense stand there is usually insufficient material in the canopy alone to permit a crown fire to spread because it must continually be supplied with heat from a rapidly burning ground fire beneath, except in hot, very dry windy weather. This fact is the key to the principal preventative measure for controlling such devastating fires—creating a discontinuity between the ground and the canopy by eliminating undergrowth and pruning the trees.

Few forest fires reach the third stage but those that do mostly occur in coniferous forest, principally due to the large amount of highly flammable dry undergrowth. Such fires are extremely dangerous to man and property as well as destructive of valuable plantations. Direct control is difficult, hence the importance of investment in protective measures that prevent fires reaching this stage. These are outlined below.

Prevention of forest fires

Prevention has both passive and active components. Measures can be taken to reduce the likelihood of a fire starting such as preventing the encroachment of moor burning with an external firebreak or reducing ground vegetation along roads and around dwellings. Other measures can be taken to cope effectively once a fire has begun. Conventionally these two aspects are considered under fire prevention and fire fighting, but the first is essential both to prevent a fire from spreading and to prepare strategic areas for fire-fighters where the intensity of fires is naturally reduced and fighting is likely to be most efficient.

Preventing fires from starting

Since most fires are started by people—either by carelessness or accident—the best method of prevention is to inform people of the risk. Unfortunately, this is only partially effective and does not work at all in the case of arson, so recourse must be made to other means of prevention. Nevertheless, an important method of reducing the fire risk is to publicize the dangers of forest fires, make the public aware of fires, especially in hazardous locations. In some instances, laws are passed that help reduce the danger such as those in Britain which limit grass burning in spring, or in Mediterranean regions which either partially or totally ban burning depending on the season, and forbid access to forests in summer.

Silvicultural methods of preventing fires starting or spreading are essential. Delabraze (1990b) gives a comprehensive description of fire-prevention oriented silvicultural principles. They generally involve three elements:

(1) increasing fuel discontinuity;

(2) design and layout of plantations;

(3) reducing fuel loadings.

It is generally preferable to have small, homogenous, dense stands where trees are the same height, thus reducing the undergrowth and favouring natural pruning. This creates a vertical fuel discontinuity. Diversification inside a stand can involve planting less susceptible species with the main crop and is an important fire protection measure in northern Europe, as for example in Lower Saxony (Otto 1982). However, it is important not to introduce slow-growing species as an understorey which may increase the vertical fuel continuity from grass through shrubs and low branches up to the crowns.

As was pointed out earlier, young crops are most at risk. If the total area of such stands in any one place can be minimized, this will help reduce the overall hazard to a forest. When laying out a plantation it is necessary to create horizontal discontinuities by designing a mosaic of stands of different species and different ages, heights, and structure. Agricultural crops can also contribute to the diversity.

In areas of high risk it is almost universal practice to construct firebreaks and rides. These are usually located along susceptible boundaries (e.g. ridges, roads, and external boundaries) and through a plantation to break up the area of forest into smaller blocks. Two types of firebreaks have two different roles:

1. Traditional firebreaks along rides and roads—although such breaks were thought to help prevent fires from spreading, their primary role is not only as a firebreak as such (reducing the risk of spread from cigarette butts or sparks from motor vehicles) but mainly to reduce the intensity of the fire, so that firefighters and equipment can be brought in to control it rapidly, with reduced risk to life. Thus, an extremely important aspect of such protective breaks is good access to them.

2. Large or 'strategic' fuel breaks—for a firebreak to stop fires spreading its width generally needs to be more than 100 m. Narrower, unplanted breaks within the forest, as well as being unsightly, may actually increase the risk of fire moving across since they act as channels for increasing wind speed and causing turbulence (Cheney 1971; Delabraze 1990b).

Because strategic fuel breaks occupy much land—at least 1 ha per 100 m length of fuel break—they are frequently not simply left as fuel-reduced zones with no other function. Instead, they can be used for grazing, left partly wooded (Fig. 13.3), or planted, wholly or partially, with trees to reduce wind speed, provide shelter for livestock, reduce grass and shrub growth, and improve the landscape.

Reducing fuel loadings is an important means of fire protection in many forests. The objective is normally to reduce the quantity of fuel—in the litter and undergrowth—to about 1 to 3 t ha^{-1}. Operations to achieve this include low pruning, as well as most of the techniques used for weed control (Chapter 8):

- cultivation—if totally weed-free areas are necessary;
- mechanical treatments—cutting or crushing;
- use of herbicides, if legislation permits—total, selective, or growth retardants, depending on the objective;
- prescribed burning;
- livestock grazing.

Fig. 13.3 Silvopastoral management for forest fire prevention in a Mediterranean *Pinus* stand. The understorey is controlled by grazing, following sward improvements. The short herb layer is less flammable than woody Mediterranean shrubs, and is maintained by grazing. (Photo: M. Etienne, INRA.)

Treatments which totally destroy all vegetation, or which allow the spread of less combustible vegetation, are usually applied along the edges of roads so that there is less fuel near where people might be, and along rides used by fire-fighters. For aesthetic reasons, as well as to reduce wind speed, widely-spaced, high-pruned trees are often left on fuel breaks. Livestock grazing inside the forest—if necessary after destruction of non-palatable species—is a current practice in New Zealand (Knowles 1991) and is increasingly used in southern Europe (Bland and Auclair 1996).

Prescribed burning inside forests is widely used in north America (Johnson 1984) and parts of Australia (Cheney 1990) and involves the controlled use of fire burning through a pole-stage or older stand to reduce the total amount of flammable undergrowth and litter to levels unlikely to burn under wild fire conditions. This practice is less common in Europe because fuel loadings are lower, and prescribed burning is mainly restricted to firebreaks along roads or tracks. It is currently used in Scandinavia (Viro 1969; Lindholm and Vasander 1987), but it is illegal in some countries, including Italy. Its use is increasing and research is being carried out in Portugal and in France (Rigolot 1993).

Very often a combination of methods can prove most efficient in limiting fire risk: mechanical cutting and crushing of the shrub layer, prescribed burning to destroy the cut material, the use of selective herbicides to slow regrowth, and livestock grazing—with pastoral improvements (oversowing or fertilizing) if necessary. A well-managed silvopastoral system can prove effective in reducing the fire risk at a lower cost than mechanical or chemical control.

Fire fighting

Effective fire fighting has two components:

(1) detection of a fire;

(2) extinguishing it.

Fire detection is primarily a question of good communication. Not only must a fire be observed but it must be reported quickly to the appropriate authority who can initiate action to suppress and extinguish it. Fire towers, other look-outs, aerial surveillance, radios, and telephones are all used to detect and report fires rapidly.

As stated previously, preventing serious fires starting is one of the main components of fire-fighting, as it not only reduces the risk but it also reduces the intensity of fires along routes intended to give rapid access to strategic areas.

Fire suppression methods

A fire will only burn if heat, oxygen, and combustible material are all present; this is the so-called fire triangle. If one element is reduced or eliminated the fire will go out. This is the basis of all fire fighting and the three main methods used against forest fires each aim to reduce a different component of the triangle.

1. Water—water suppresses fires by lowering the temperature because much heat is needed to vaporize the water. It is applied from back packs, portable pumps, direct from vehicles, or from the air. A supply must be available to replenish water-dispensing units such as roadside dams, special tankers, perennial streams, etc. Additives are often used to increase the effectiveness of water and they fall into two categories. The first increases the viscosity of the water to enable more to be held on tree foliage. Alginates are often used to achieve this. Foams have a similar effect and are more visible. The second group of additives are wetting agents, or surfactants, which lower surface tension to overcome the naturally water-repellent characteristics of most leaves and needles. Fire retardants provide another way of reducing the heat of fires of low intensity, by interfering with both the physical and chemical processes of combustion. They generate non-flammable gases and promote glowing rather than flaming and hence reduce the heat output and the rate of spread of a fire. The most widely used retardants are ammonium sulphate and diammonium phosphate. Both are used as fertilizers as well, and are cheap, easily available, and do little environmental damage.

2. Beaters—the fire is hit with a flat surface which temporarily excludes oxygen. There are many types of beater but all are basically a handle with a flap on the end, indeed even *Betula* branches or young trees can be used. Beaters are only suitable for surface fires which are safe to approach within about 1 m. Digging and throwing soil on to a fire fulfils a similar role.

3. Reducing fuel—sometimes it is possible to clear combustible material in the path of a fire by bulldozing a break or counter-firing. This may be the only feasible suppression method. However, counter-firing is very difficult to implement, and can cause legal problems: voluntarily lighting a fire is forbidden by law during the periods of high fire risk, and lighting a fire on a private property not yet touched by a wildfire can be construed as arson.

A more complete discussion of the main ways of combating fires is given by Pyne (1984) and Delabraze (1990*a*).

Part III Specialized forms of plantation silviculture

14 Plantations on disturbed land

The artificial substrate and dereliction

A consequence of many mining and other industrial activities is the formation of disturbed land, often consisting of structureless material largely devoid of topsoil and subsoil (e.g. Fig. 14.1). Much of this land is 'derelict' and is defined as 'land so damaged by industrial or other development that it is incapable of beneficial use without treatment' (Wickens *et al*. 1995). Many such sites, epitomized by the colliery spoil heap in industrialized regions, are frequently ugly, may erode seriously, and become an environmental hazard. Quite apart from the loss of productive potential of the ground which has been spoilt, these other factors make rapid restoration an important priority. Experience has shown that the cost of restoring much of this kind of derelict land to agriculture is prohibitively expensive. Thus restoration, which used to be viewed as no more than the land being of an acceptable shape and appearing green, was frequently 'solved' by planting trees. This brought more land under trees, especially in industrialized regions which often lacked tree cover. However, foresters should not welcome unreservedly the prospect of such planting and today objectives of sustainable development require greater attention to soil-forming materials and processes as an integral part of reclamation. Derelict sites bring many problems, and even if establishment of trees is satisfactory on many kinds of sites, there is still considerable uncertainty about whether a productive crop can be grown successfully.

Numerous industrial processes disturb land of which the principal ones are mining for coal and various ores, extraction of sand, gravel, and clay (including china clay—kaolin), rock and limestone quarries, and the deposition of waste products such as pulverized fuel ash and domestic refuse in landfill sites. The scale of disturbance relates not only to the kind of product being won but the way in which this is achieved. Deep mining yields tips and spoil heaps, opencast working in strip or surface mining involves removal and redistribution of large quantities of overburden, and accumulation of tailings in lagoons follows many metal extraction processes.

The greatest amount of dereliction occurs in industrial countries. Although usually less than half of 1 per cent of all land is degraded in this way, dereliction tends to occur in regions of high population, thus magnifying the despoilation. As an example of the areas involved, Table 14.1 presents estimates for different regions of England.

Fig. 14.1 A site shortly to be reclaimed after bauxite mining at Abbaye de Thoronet, France. (Photo: P. Allemand, INRA.)

Table 14.1 Location and types of derelict land in England in April 1993 (Source: Wickens *et al.* 1995)

	Inner City (ha)	Other Urban (ha)	Rural (ha)	Total (ha)
Spoil heaps	435	3518	5237	9191
colliery spoil heaps	219	1930	1960	4109
metalliferous spoil heaps	8	523	2473	3003
other spoil heaps	209	1065	805	2079
Excavation and pits	195	1476	4135	5807
Military dereliction	77	397	2801	3275
Derelict railway land	635	2199	2782	5615
Mining subsidence	88	425	162	674
General industrial dereliction	2904	4657	2188	9749
Other forms of dereliction	909	2565	1815	5289
Total	5243	15 236	19 121	39 600

Problems

The main problems on different types of disturbed sites and the means of restoring a cover of vegetation have been described in detail by Moffat and McNeill (1994),

Bradshaw and Chadwick (1980), and Fox (1984). The problems arise principally because the substrate to be reclaimed is almost always derived from mining or earth moving, and is largely undeveloped subsoil or rock. That such conditions are inhospitable to plant growth is indicated by a virtual absence of natural colonization on many wastes (Fig. 14.2). This phenomenon is not exclusive to artificial substrates and can be found associated with volcanic lavas and ashes, sand dunes, and along the course of retreating glaciers where development of soil structure may take up to 100 years. Similarly inhospitable conditions are also found in many parts of southern Europe where many centuries of overgrazing and burning have led to the loss of tree cover, subsequent erosion, and the loss of soil. Costly reclamation schemes are being carried out in many places (Fig. 14.3). Time will eventually solve most of the problems, unless there is extreme toxicity, as the natural soil-forming processes take place, but the number of decades needed are often unacceptabe unless natural colonization is itself a desired aim of land management.

The problems confronting successful afforestation are outlined below and fall into three main categories; those of soil conditions, exposure, and protection.

Physical characteristics of the rooting medium

The physical conditions of many substrates arising from industrial processes are inhospitable for tree growth.

Fig. 14.2   A reclaimed opencast coal-mining site in the south Wales coalfield, planted with larch (*Larix kaempferi*) 4 years previously. Note the almost complete lack of colonization by weeds of this shaley site. (Crown Copyright.)

Fig. 14.3 Land restored to forest after centuries of overgrazing and burning in the Alps of Haute Provence, France. (Photo: F. Paillard, INRA.)

Structure and texture

Unless soil is stripped, stored, and replaced, the substrate is usually ground rock fragments, clay, or hard rock with no organic matter and little biological activity to aid aggregation into a satisfactory structure. Such material, often made up of large solid particles, holds little water and where rainfall is intermittent, combined with the exposed condition of most sites, leads to rapid drying of the surface and consequent drought problems for newly planted trees. Another consequence of poor substrate structure and texture is the development of high temperatures at the surface, especially if the material is dark in colour. Lack of protective vegetation and a rapidly drying surface can result in lethal surface temperatures, in excess of 50°C on hot days.

In addition to these unfavourable inherent properties of many substrates is the effect of compaction caused by bulldozers and box-scrapers that are often used in the early stages of restoration. The substrate easily becomes compacted and consolidated and so presents barriers to root growth and drainage. Compacted soils suffer poor aeration and water cannot infiltrate or drain readily, and thus both drought and waterlogging are a feature. Compaction affects not only the structure. When topsoil is stripped and stored, often for years while mining is in progress, and then respread on the surface, it may not behave immediately as if it had never been removed. Storage in large heaps results in compaction, the development of anaerobic conditions, and the consequent loss of many of the organisms, such as earthworms, which aid nutrient cycling and plant growth. Alleviation of compaction is an important stage in all reclamation work.

Drainage

The nature of the substrate, compaction, and poor reshaping can all lead to problems of drainage. Unless excess water can drain from a site, conditions for tree growth will be poor. However, the problem is not solved simply by ensuring adequate slopes and appropriately placed water courses for two reasons:

(1) many substrates are easily erodible and rapid shedding of water can be a major hazard;

(2) most water draining from disturbed land can have a high sediment content and cannot be fed directly into the natural drainage network of streams and rivers outside the site without first flowing into siltation ponds.

Stability

The physical composition of many waste materials, combined with the lack of surface vegetation, exposes them to erosion by wind and rain and the risk of landslips. The problem is worst on steep-sided tips and heaps but with some substrates even a slope of only 5° may slip under conditions of prolonged heavy rain as the material becomes more fluid. Careful consideration of this in a restoration plan is essential.

Chemical characteristics of the soil

Nutrient supply

Both the lack of organic matter and the unweathered state of much waste material can lead to nutrient supply problems. By far the commonest is a shortage of nitrogen: the total nitrogen content may be as little as 100 kg ha^{-1} of nitrogen compared with figures of 750 to 1500 kg ha^{-1} in normal topsoil. In addition to inherent nutrient deficiencies for plant growth is the lack of any effective pathway to aid nutrient cycling. Thus, until organic matter levels increase substantially, the successful establishment of trees depends on applying organic wastes, and to aid nitrogen enrichment, interplanting trees with legumes and/or using nitrogen-fixing species such as *Alnus* or *Robinia*. The use of species which facilitate nutrient cycling, such as the deciduous *Larix* and broadleaved pioneers such as *Betula* and *Salix,* is also important.

Toxicity

Some mining wastes contain such high concentrations of elements that they are toxic to plants and soil organisms; notably copper, zinc, and lead. Even small amounts can be sufficient to prevent colonization while leaching of these elements can be a serious pollution hazard to rivers. Problems are worst on older mined wastes because methods of extraction were less efficient, leaving more of the metal in the waste material. In Britain, there are copper workings dating from Roman times where vegetation development is still sparse, and there are many deposits and areas of land down wind from chimneys and flues polluted in the 1800s on which hardly any plant growth has occurred.

Acidity and alkalinity

The presence of pyrite (FeS_2) in some wastes, and especially the colliery spoils from deep mining, causes acidity as it weathers by releasing sulphuric acid. Low pHs of between 2 and 4 are inhospitable to most plants as well as leading to pollution of streams and rivers. The extent of the problem depends on quantity and particle size of the pyrite, but in serious cases it is either necessary to wait many decades until the pyrite has mostly weathered and the acid has leached from the soil or to modify reclamation practices and avoid depositing pyritic spoil as the final cover (Moffat and McNeill 1994). It is feasible to apply lime but high quantities may be needed ($100-300$ t ha^{-1}) which create their own problems.

A few industrial wastes give rise to strongly alkaline substrates of pH 9 or more, for example pulverized fuel ash (PFA) and the waste from some soda ash processes. Once deposited, weathering takes place which will eventually lower pH to about 8, a level more amenable to revegetation.

Organic matter

Nearly all substrates from industrial mining processes have very little or no organic matter. As has been mentioned, this affects the biological, physical, and chemical characteristics of the substrate.

Exposure

Trees planted on disturbed land are frequently exposed both to wind damage and to airborne pollution.

Wind

The lack of vegetation on many sites, and often the nature of the local topography on reclaimed sites, usually exposes the planted trees. As well as impairing growth and leading to poor form, exposure causes two direct problems. First, a common phenomenon is for young trees to rock in the planting holes. Where the substrate consists of large, angular particles and stones such rocking or swaying in the wind can lead to abrasion and damage to the bark and cambium at the root collar. Secondly, the dry surface, so characteristic of many wastes, is readily eroded by wind exposing the roots, while the fine gravel and sand whipped up will abrade the young trees.

Airborne pollution

Most reclaimed sites are in large industrial areas where atmospheric pollution is a common problem. Many trees are more susceptible to airborne pollution than annual plants. In the past sulphur dioxide emission from various industrial processes, such as aluminium smelting, was a common pollutant damaging trees downwind. Various kinds of pollution, including ozone episodes, may further stress trees used in rehabilitation and affect the choice of species.

Protection

Mammal damage

Trees planted on reclaimed land are particularly susceptible to damage from rabbits, hares, and livestock because there is very little other vegetation to browse. If such animals get on to the planting site their only food, often for many hectares of land, is the trees planted at regular intervals. Browsing by a few animals can devastate a young plantation in a few hours. Even if a low vegetation cover has been established as part of the reclamation process, planted trees are often vulnerable.

Fire

Although very young plantations on reclaimed sites are not particularly susceptible to fire damage because vegetation is sparse, once they have reached the thicket stage fire becomes a serious hazard. This problem is exacerbated because, as has been mentioned, many such sites are in industrial regions and areas of high population. Thus, not only is the crop itself a hazard but the risk of a fire starting is high.

Restoration of sites for tree planting

Restoration plan

Until the 1970s little thought was given to the restoration of a tip or opencast mine as it was being worked. This frequently led to the most inhospitable environments with which to work. Happily this is rarely the case today and plans for satisfactory restoration are prepared as part of the whole operation. Thus, the organization undertaking the extractive mining or spoil deposition becomes responsible for the condition of the site afterwards and it is very much in its interests to leave the ground reasonably shaped and as suitable for revegetation as is practicable

(Fig. 14.4). Agreeing the need to conserve topsoil, identifying and storing separately soil forming materials as the overburden is removed, planning site drainage, avoiding high walls where cliffs are left exposed, and ripping compacted surfaces can all be included in the plan.

Site preparation

All site preparation has three aims:

(1) to leave a safe and stable substrate;

(2) to shape the restored ground in ways which blend with the landscape;

(3) to make the site hospitable for trees.

   Moffat and McNeill (1994) deal with each of these aims and describe the steps in reclaiming land to forestry. This section is principally concerned with the third, that is how to make a site hospitable to trees.

Retaining topsoil

The difficulty of revegetating much disturbed land is caused by the extremely unfavourable physical characteristics of the substrate. This is most simply and easily overcome by respreading a layer of topsoil or subsoil, once reshaping operations are

Fig. 14.4 The entire area in the photograph is restored land following opencast coal mining in the upper Neath valley, south Wales. Note gentle slopes, sympathetic landforms, and ripped cultivation lines ready for planting. Fencing is essential to prevent sheep browsing of young trees.

finished. Ideally soil should be respread to a depth of 50 cm or more. Below this should be layers of soil forming materials, such as weathered shales, which possess minimum characteristics of bulk density, chemical, and physical properties (see Moffat 1987). The most obvious source of soil for respreading is from the ground before it is worked. An important part of the operational plan of any mining activity should be to provide for removal and storage of topsoil, subsoil, and soil forming materials for subsequent respreading when restoration begins.

Microtopography

It is clear that the physical condition of many substrates can lead rapidly to extremes of drought or waterlogging. This can be significantly alleviated by ensuring that there are no flat areas or large depressions where water will accumulate. Gently rolling ground with slopes of 3 to 5° is ideal.

Research has demonstrated the value of shaping the ground into a system of ridges about 30 m wide (Fig. 14.5) and with the top 1 to 1.5 m above the intervening furrow (Binns and Crowther 1983). The line of the ridge itself should also slope gently. Such landforms alleviate drainage problems and promote tree growth except in drier regions where a deeper topsoil layer may be needed (Moffat and Roberts 1989).

Preventing compaction

Restoration of a site to visually acceptable landscape form and creation of the appropriate microtopography usually necessitates heavy earth-moving machinery. Apart from loose tipping practices, this work inevitably leads to compaction, a condition

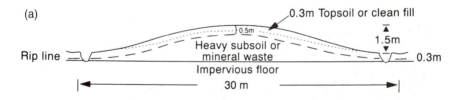

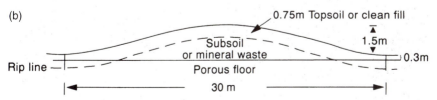

Fig. 14.5 Ridge and furrow landforms: (a) a system for sites with high water tables; (b) system suitable for permeable sites. (From Moffat and McNeil 1994. Crown Copyright.)

which must be alleviated before planting. This is achieved by deep tining: the use of a winged tine to lift the soil slightly as the tine moves through it has proved effective.

Improving nutrition

Application of fertilizers may be necessary, especially nitrogen. Where the latter is seriously deficient, annual or even more frequent, applications may be needed with some tree species for a few years. Heavy applications of organic wastes, such as sewage sludge, are also useful for improving the organic matter content and providing nutrients. Overcoming problems of severe acidity are solved partially by heavy applications of lime if it can be well mixed to a depth of 50 cm or more. However, this needs careful management as applications in excess of 100 t ha^{-1} may upset the calcium:magnesium balance of the spoil and restrict phosphate uptake (Costigan *et al.* 1982).

Choice of species

The nature of reclaimed sites necessitates the use of species which are tolerant of exposure and undemanding nutritionally. Such characteristics are those associated with pioneer species and it is not surprising to find that they are usually chosen for planting. *Alnus* and *Robinia pseudoacacia* are outstandingly valuable in most temperate climates as they fix nitrogen and tolerate moderate compaction as well as acidities down to pH 3.5.

The choice of tree species is important and investment in other preparatory measures will be devalued if care is not taken in matching species to the exacting conditions reclaimed land so often presents. Table 14.2 from Moffat and McNeill (1994) lists potential species based on British experience.

Establishment

Supplementary vegetation

It is common to sow a sward for ground cover, to help build the soil and sometimes to prevent surface erosion. Among grasses, *Festuca ovina* and *Agrostis tenuis* are favoured as they are not too competitive with the trees. Legumes include relatively unpalatable species such as everlasting peas (*Lathyrus sylvestris* and *L. latifolius*), white clover (*Trifolium repens*), and tree lupins (*Lupinus arboreus*) and are valuable pioneer plants.

Planting

Though direct seeding has been successful where conditions are suitable, it is most common to establish trees by planting transplants or container grown stock. The

Table 14.2 Suitable tree species for reclaimed sites in Britain (based on Moffatt and McNeill). Species are classed as tolerant (□□), moderately tolerant (□), or intolerant (x) to heavy soils (likely to be seasonally waterlogged), calcareous soils, acidic soils, exposure and air pollution.

Species	Heavy soils	Calcareous soils	Acidic soils	Exposure	Air pollution	Comments
Broadleaves						
Acer campestre	□	□□	□	□	□	
Acer platanoides	□	□□	x	□□	□	
Acer pseudoplatanus	□	□□	□	□□	□□	
Alnus cordata	□	□□	x	x	□□	Nitrogen-fixing
Alnus glutinosa	□□	□	□	□□	□□	Nitrogen-fixing
Alnus incana	□□	x	□	□□	□□	Nitrogen-fixing
Alnus rubra	□□	x	□	□□	□□	Nitrogen-fixing
Betula pendula	x	x	□□	□□	□	
Betula pubescens	□	x	□□	□□	□	
Crataegus monogyna	□	□	□□	□□	□□	
Fraxinus excelsior	x	□□	x	x	x	Fertile sites only
Populus alba	□□	x	x	□□	□□	
Populus canescens	□□	□□	□	□□	□□	
Prunus avium	x	□	x	x	□	
Quercus robur	□	□	□□	□	□□	Fertile sites only
Quercus rubra	□	□	□□	□□	□□	Fertile sites only
Robinia pseudoacacia	□	□	□□	x	□□	Nitrogen-fixing
Salix caprea	□	□	□□	x	□□	
Salix fragilis	□□	□□	x	x	□	

Table 14.2 *continued*

Species	Heavy soils	Calcareous soils	Acidic soils	Exposure	Air pollution	Comments
Broadleaves						
Sorbus aria	□□	□□	□□	□	□	
Sorbus aucuparia	□	□	□	□□	□	
Sorbus intermedia	□□	□	□	□	□	
Conifers						
Larix decidua	□	×	□	□	×	
Larix kaempferi	□□	×	□	□	□	
Picea sitchensis	□	□	□□	□□	×	Fertile sites only
Pinus contorta	□□	×	□□	□□	×	North of UK only
Pinus nigra subsp. laricio	□	□□	□□	□□	□□	Below 250 m in UK
Pinus sylvestris	×	×	□□	□□	×	

principles do not differ from the establishment of any other crop but because of the more difficult site conditions it is more important that the plants are healthy and have a good store of nutrients, and that they suffer no desiccation or other damage between lifting from the nursery and planting. Particular attention must also be paid to replacing dead plants which are often numerous on disturbed land, good weed control, and protection from animals.

15   Short-rotation crops

In temperate regions, short-rotation crops may be considered as those grown on rotations of less than about 30 years. Many traditional, broadleaved coppice crops fulfil this definition but it also includes some short-rotation plantations grown for specialized products, notably *Populus* spp. and *Eucalyptus* spp. and some conifers grown directly for pulp with no intermediate thinnings. With the present European trend towards diversified land uses, and in particular the 'set-aside' policy, increasing areas of former agricultural land will become available for non-food crops. Rather than having this land left fallow, one interesting possibility for farmers is short-rotation forestry. In addition, since the 1970s' energy crisis and with increasing environmental concern about energy production—whether from carbon dioxide-producing fossil fuels or from nuclear plants—several alternative energy sources are being considered, including short-rotation forest biomass (Christersson 1994).

Traditional coppice

The production of firewood, poles, and many other types of small- and medium-sized material on short (10 to 30 year) cutting cycles is one of the oldest forms of management in semi-natural forests. Production is achieved by coppicing and reproduction from stool shoots of many species such as *Quercus* spp., *Fraxinus* spp., *Castanea sativa*, *Corylus avellana*, *Populus* spp., and *Salix* spp. Nearly all currently worked coppice crops were originally planted, though, in the past, naturally occurring mixed woodland was often coppiced for firewood, primitive building materials, fencing, etc. Coppices of *Castanea sativa* were planted for firewood and split fencing, osiers for basket-making, and *Corylus avellana* for hurdles (Crowther and Evans 1986).

In much of Europe, the demand for traditional coppice produce declined rapidly from about 1870 until, by the early twentieth century, it almost ceased to exist. The decline started with the industrial revolution when new technology and inventions made available cheaper and better alternatives to the traditional forest produce. For example wire replaced *Corylus avellana* hurdles for fencing and the efficient transport provided by new railways enabled coal to be taken to the countryside, replacing fuelwood. The provision of electricity to rural areas in the 1920s and 1930s caused the final decline, with some increases in use due to war-time energy shortages. However, large areas are still managed as coppice or coppice with standards, especially in France (5 Mha) and Italy (3.7 Mha).

Coppice can have higher mean annual increments than conventional forest crops on comparable sites, on a total woody biomass basis. But, due to the fact that timber produced from coppice is of lower quality than that of high forest, coppice has most often been left on the poorer soils, better sites being devoted to high quality conventional forest—as indeed even better sites are dedicated to agricultural production. The comparison between yields is therefore often to the detriment of coppice (Auclair and Cabanettes 1987; Cabanettes 1987; Bergez *et al.* 1989). However, for some purposes coppice can be more attractive than conventional forestry. For example:

1. The material can be reasonably easily handled by inexpensive machinery which can often be attached to a farm tractor. The costly and very specialized equipment for dealing with large trees is not needed.

2. A usable and marketable product is available in a short time. This has many attractions to landowners (farmers in particular) who are not prepared, or cannot afford, to invest in long time scales of normal forestry.

3. Coppice crops can be grown on land which is too wet to grow most agricultural crops, such as surface water gleys. These conditions make arable cropping difficult and the successful production and use of grass depends upon high levels of management. The use of heavy logging equipment is also impossible and small machinery must be used.

4. Coppice produce air-dries quickly because of its small dimensions. This makes it much more quickly available for use than larger produce.

5. Re-establishment costs following harvesting are very small compared with costs of replanting a conventional tree crop, and in most cases they do not exist at all.

Short-rotation coppice

The energy crisis of the early 1970s stimulated renewed interest in short-rotation crops of coppice in temperate countries. This has resulted in a remarkable expansion of work and literature on the subject which was reviewed by Mitchell *et al.* (1992). Planted crops which are subsequently coppiced on 1- to 5-year cycles are economically valued as alternative sources of wood, charcoal, and liquid fuels, a basis for chemical processes, wood pulp, and sometimes as a fodder supplement. In non-industrial societies, mostly in the tropics, coppice material is used as fuel for cooking and heating following widespread deforestation. In many developing countries up to 90 per cent of the total energy used comes from wood.

The main interest in these crops is in northern European countries such as Finland, Sweden, and Ireland where there are no worthwhile supplies of fossil energy resources, making it necessary to import all conventional energy fuels. In 1993, almost 10 000 ha of farmland in southern and central Sweden were planted as 'energy forests', producing an average yield of 10 to 12 t ha^{-1} year^{-1} of dry matter in 3- to 5-year rotations (Christersson *et al.* 1993). Forest crops, of course, are not the only sources of biomass for energy, though they are among the most efficient in terms of the ratios of energy contained in the harvested crop to total energy input (Table 15.1). Field crops such as stubble turnips, rape, and sugar beet, crop residues, and animal wastes can also be used for energy production. In Brazil, sugar cane and cassava produce a significant amount of ethanol, used as a fuel for motor vehicles (Monaco 1983).

Species

Broadleaved, rather than coniferous, species are generally planted for short-rotation crops for two reasons:

1. Costs can be reduced if more than one harvest can be made from one establishment operation. Hence one of the most important attributes of a species is that it should coppice vigorously. Many broadleaved species coppice well but few conifers do so at all, one exception being *Sequoia sempervirens*. Rapid coppice growth is attributed to the fact that roots already exploit the soil fully and also contain the carbohydrate reserves which allow the growth (Auclair *et al.* 1988; Dubroca 1983).

2. Unlike evergreen conifers, deciduous broadleaved trees invest little assimilated material in their leaves and the leaves are, per unit area, photosynthetically

Table 15.1 Maximum yields and energy contents of various crops (adapted from Hall 1983)

Crop	Maximum yield (dry weight) (t ha^{-1} yr^{-1})	Energy content of cultures (GJ ha^{-1})[*]	Cultural energy ratio[+]
Conventional forestry	15	225	10–20
Short-rotation forestry	12	180	5–15
Algae	60	900	8 to high
Catchcrops	8	120	3–4
Grass	15	225	2.4–5.6
Wheat	5	75	3.4
Sugar beet	10	150	3.6

[*] 1 tonne oil equivalent = 42 GJ.
[+] i.e. harvested crop energy to total energy input.

more efficient than those of conifers. Hence a larger amount of assimilate is available from an early age for the growth of stems, branches, and roots. Though evergreen conifers may eventually grow faster, they take many years and a large amount of assimilated carbon to build up an efficient canopy.

Paradoxically, the most favoured trees for short-rotation crops are often species, or close relations of species, which are regarded as weeds in long rotation plantations. They are often pioneers of natural successions which establish quickly and have vigorous juvenile growth. Short-rotation crops of *Salix* are increasing with the development of 'energy forestry' in Scandinavia and in Ireland. The preferred planting stocks are clones of *Salix viminalis* and *S. dasyclados*. They are managed in a very similar way to conventional agricultural crops (Sennerby-Forsse and Johansson 1989; Christersson *et al.* 1993). Species and clones of *Populus* (Fig. 15.1) are also commonly planted but *Platanus* and *Eucalyptus* are increasingly being used, the latter especially in northern Spain and Portugal where many thousands of hectares have been established in recent years for the production of pulpwood. Each clearly has a preferred range of sites on which growth is rapid. There is increasing interest in nitrogen-fixing species such as *Alnus* and *Robinia* though

Fig. 15.1 Two-year-old poplar shoots showing potential for biomass production. (Crown Copyright.)

yields from these are not expected to be as high as from other species because the process of nitrogen fixation consumes quantities of assimilated carbon. Their main interest is their ability to grow on relatively poor soils.

In most cases, the clones used at present are those which have been selected for high-volume production and disease resistance in conventional stands rather than in coppice production.

Yields of short-rotation crops

From the first experimental results in the 1970s, Cannell and Smith (1980) and Pardé (1980) suggested that in most temperate regions it would not be realistic to give field predictions higher than about 6 to 8 t ha^{-1} year^{-1} of wood dry weight in stems and branches. Such levels, defined by Hansen (1988) as 'field yields' as opposed to 'record yields' obtained in very tightly controlled conditions, are remarkably consistent between different species and regions. They compare with yields of 4 to 7 t ha^{-1} year^{-1} from the traditional but very productive, (by standards in most of Europe), *Castanea sativa* coppice in southern England and central France (Evans 1982*b*; Auclair and Cabanettes 1987; Cabanettes 1987) (Fig. 15.2).

Yields up to 10 to 12 t ha^{-1} year^{-1} have now been confirmed in practice in the field, for intensive short-rotation coppice plantations in Sweden (Christersson *et al.*

Fig. 15.2 Recently felled sweet chestnut (*Castanea sativa*) coppice with 14-year-old coppice crop in background. (Crown Copyright.)

1993). These compare with averages of about 10 to 11 t ha^{-1} year^{-1} for productive conifers such as *Picea abies* on long rotations, and more than 15 t ha^{-1} year^{-1} from fast-growing conifers such as *Pseudotsuga menziesii* on good sites.

One of the main differences between traditional coppice and modern, high yielding short-rotation coppice concerns the level of input, or intensification: the latter requires regular fertilizer, herbicide, and pesticide treatments to maintain the high yields. Intensive short-rotation coppice is a high cost–high return system, compared to the low cost–low return system in most traditional existing coppice (Auclair and Bouvarel 1992), and economic considerations must be examined seriously (p. 226). More extensive *Salix* plantations in Sweden are reported to yield 6 to 8 t ha^{-1} year^{-1} when no fertilizer is added (Christersson *et al.* 1993).

Higher figures, exceeding 20 t ha^{-1} year^{-1}, have been obtained in optimum experimental situations, but they should not be extrapolated to field conditions without extreme caution. Potential levels of biomass production as high as 25 t ha^{-1} year^{-1} are thought possible by Hall (1983) through application of site-specific research into crop and tree performance using selected species and clones adapted to varying geographic and climatic conditions. Fundamental investigations into aspects of photosynthesis are also needed. But even the most optimistic estimates of future field average production do not exceed 15 t ha^{-1} year^{-1} (Christersson *et al.* 1993).

Shortening the rotations does not necessarily increase average annual biomass yields. Cannell (1980) stated that vigorous *Populus* clones, *Salix,* and *Nothofagus* will give mean annual increments of 6 to 8 t ha^{-1} year^{-1} dry weight on good sites in 1 year if planted at 250 000 stems ha^{-1} or 25 years if planted at 2000 stems ha^{-1}. Auclair and Bouvarel (1992) showed that *Populus* hybrid short-rotation coppice could produce equal annual biomass yields in 1-year rotations at 20 000 stems ha^{-1}, in 2-year rotations at 10 000 stems ha^{-1}, or in 3-year rotations at 5000 stems ha^{-1}, over a period of at least 6 years.

Silviculture

To achieve reasonably high levels of production, good soils are needed, which often brings energy crops into direct competition with arable agriculture. Most research outside Ireland and France has been done on arable soils. The land must be reasonably flat for mechanized working and the soils must have a good structure and texture, adequate water, and sufficient nutrients and organic matter. Most of the technical problems on such sites have been solved. Some peatlands may also be suitable but they pose problems of drainage, cultivation, and nutrition.

The initial establishment of short-rotation crops is intensive and costly. Planting densities are high and are adjusted according to the expected length of rotation and size to which the crop is to be grown before harvesting. Because rotations are short it is important that the available space is occupied quickly, so much greater numbers are planted than in long-rotation crops. Usually no fewer than about 5000 stools ha^{-1} are needed and more commonly between 10 000 and 20 000 for 3- to 5-year *Populus* and *Salix* rotations (McElroy 1981; Sennerby-Forsse and Johansson 1989). With traditional *Castanea sativa* on a 12- to 15-year cycle (Fig. 15.2), the aim is for 800 to 1100 stools ha^{-1} (Crowther and Evans 1984).

Plantations of selected *Populus* and *Salix* clones are usually established as cuttings, and thorough site cultivation and weed control are essential. Adequate protection by fencing and pest and disease management is necessary. Because most of the total above-ground biomass is harvested, the nutrient cycle is interrupted. Even if the leaves are not harvested, a large amount of mineral nutrients is concentrated in the smallest parts of the trees—twigs and bark—which are removed from the site (Ranger *et al*. 1986, 1988). Fertilizers, especially nitrogen, must often be applied to replace nutrients removed in the wood to maintain yields. Applications of nitrogen of 80 to 120 kg ha^{-1} year^{-1} are often recommended (Faber and van den Burg 1982; Sennerby-Forsse and Johansson 1989), or irrigation with waste water or liquid animal manure (Christersson *et al*. 1993). Irrigation to prevent water stress can also improve yields. Mechanical harvesting is usually carried out in winter, with heavy machinery which, if the ground is not frozen, may compact the soil and result in poor stool recovery and subsequent growth.

Although the early growth rate of coppice shoots is much greater than that of seedlings or cuttings, the extent to which production levels of first rotations can be maintained in future harvests has been questioned (Pardé 1980; Wright 1988). Auclair and Bouvarel (1992) have shown that after six 1-year rotations hybrid poplar coppice could maintain regular yields—depending on climatic conditions. Possible reductions in yield would be due mainly to nutrient losses rather than physiological stresses. There is little information about the longevity of stumps, which appears to be species-specific: *Castanea sativa* can be coppiced for many cycles, whereas stump mortality of *Betula* is reported to increase after three rotations. This can be related to the ability of the former species to produce new roots at each coppice rotation (Bédéneau and Auclair 1989).

Short-rotation crops, like any intensively managed monoculture, are likely to reduce the diversity of the ecosystem but less-intensive, longer-rotation more traditional coppice systems have the opposite effect and are much favoured by conservationists (Rackham 1976; Peterken 1981, 1996).

Uses of coppice material

Traditional coppice has long been a main source of firewood—and still is in areas where it is abundant. Today it is also a source of logs for open fires, combining amenity and heating. Other uses of coppice are poles and fencing materials, baskets (*Salix* spp.), and some local traditional crafts such as wine barrels or walking sticks. The main economic interest in short-rotation crops is their potential for conversion into useful energy. The most obvious way of doing this is by burning to produce heat. Woodfuel has recently become a practical alternative source of energy for homes—mainly community housing—and some industrial applications in several areas which formerly relied on fossil fuels. This process has been helped by the development of automatic feeding systems which transport wood chips directly to very efficient boilers capable of using freshly harvested wood. Charcoal, gas, and liquid fuels such as methanol can also be produced (Carre and Lemasle 1987). Commercial applications for many purposes are reasonably well developed. Stassen (1982) described the combination of gasifiers and internal combustion engines for generating electricity.

Salix chips have been converted into feeding nuts for cattle (McElroy 1981), and *Populus* into hardwood pulp (Cannell 1980), both with indifferent success. Many other processes and applications of short-rotation crops are described by Grassi *et al.* (1987) and Hummel *et al.* (1988).

Economic considerations

The considerable interest in experimenting and developing technologies for short-rotation energy crops has not, so far, been matched by any significant industrial applications in most temperate countries. It seems unlikely that this will happen until the costs of energy from other sources, particularly oil, rise considerably and so cause short-rotation wood crops to become at least equivalent in value to wood produced for other purposes. In many places costs of production are very high in relation to revenues in comparison with, for example, oil, natural gas, and coal. Often the costs of harvesting fuelwood, including transport, are greater than the value of the material itself so that only locally can it sometimes be competitive. The same problem arises with some other residues, such as straw. In addition, in crowded countries like Britain, the provision of a significant area of land would present problems and cause conflict.

Only in countries where political initiatives have been taken has any real progress been made. In Sweden, Koster (1981) recorded that a start was to be made in 1985 on replacing 20 per cent of the country's oil imports by planting 100 000 ha with short-rotation crops. The European Parliament's view in September 1982 was that it would be desirable to divert some farmland to energy crops rather than continue producing surplus milk, grain, and wine (Seligman 1983).

In most places in temperate regions, coppices are likely to play only a minor role in most national energy strategies in the immediate future. In the longer term, they may become important. Current research and development will ensure their eventual success and, unlike most tree crops, it will be possible to create productive coppices in a short span of time.

Poplar plantations

There are a little over 30 species of *Populus*. They occur throughout the northern hemisphere in most of the boreal and temperate zones between the subarctic and subtropical regions. Most *Populus* spp. are planted for screening, shelter, and for production of matchwood and chip baskets or vegetable crates. *Populus* plantations (Fig. 15.3) cover more than 0.5 Mha in Europe, and are increasing at a rate of 1 per cent a year (Valadon 1996). A comprehensive account of *Populus* cultivation is in FAO (1980) and in Commission Nationale du Peuplier (1995).

Three main species are present in Europe: the Eurasian black poplar (*P. nigra*), and the north American *P. deltoides* and black cottonwood (*P. trichocarpa*). A number of cultivars are hybrids between these species: hybrid black poplars (*P. × euramericana*) derived from *P. deltoides × P. nigra* crosses, or interamerican poplars (*P. × interamericana*) derived from crosses between *P. deltoides* and *P. trichocarpa*. Important research programmes are under way to improve the genetic stock, mainly for resistance to the bacterial canker *Xanthomonas populi* and to *Melampsora* rusts. New *P. alba × tremula* clones are being developed for their adaptation to soils which are temporarily waterlogged (Lefèvre *et al.* 1994).

Although *Populus* spp. will grow on a wide range of sites, to achieve the rapid growth for which they are well known, sheltered sites with base-rich loamy soils where the water table is within about 1.0 to 1.5 m of the surface are required.

Fig. 15.3   Belgian poplars at 9 years old.

Almost all *Populus* spp. are raised by vegetative propagation either by growing forest plants from hardwood cuttings 10 to 25 cm long or by inserting longer cuttings, called setts, which are usually 1-year unrooted shoots about 2 to 3 m long. Setts are directly inserted into deep holes at the final planting position.

Populus spp. are strongly affected by competition with either weeds or each other, and to obtain fast individual growth are planted at very wide spacings, generally ranging from 5×5 m to as much as 8×8 m. In many plantations no thinning is carried out. On the best European sites veneer-sized logs, for peeling for matchwood or vegetable crates, are reached in 12 to 15 years depending on site quality. High pruning is essential for veneer quality and it is normally done to 6 m early in the life of the crop. Pruning continues regularly at 2-year intervals throughout the rotation to minimize epicormic growth.

16 Plantations for special purposes

Most plantations are established for the production of wood that can be used for a variety of industrial purposes, though some are grown for a particular specialized use, such as *Populus* spp. for the match-making and vegetable crate industries, or for energy. These are discussed in Chapter 15. But, as pointed out in Chapter 1, trees are also planted for many other purposes. The economic values of produce other than wood and the environmental benefits of trees and forests can sometimes be more important than timber. Examples of non-wood produce, conventionally called minor forest products, are seed crops from sweet chestnut (*Castanea sativa*), walnut (*Juglans regia*), and stone pine (*Pinus pinea*), or exudates such as resins and maple syrup, the bark of *Quercus suber* for cork, and that of other species for tannins, and a variety of medicinal and industrial compounds. Products of animals which feed on certain trees may have a considerable economic value including silk, lac, and honey. Fungal associates of specific trees, such as truffles, may be prized for food, and many plants of the forest understorey produce edible berries or substances used in perfume or medicinal industries. Trees are planted for the ornamental value of their foliage, for example species exhibiting rich autumn colouring, and also for the fodder value of their leaves and fruits. They are planted for environmental reasons which can be difficult to value in economic terms: for beauty, shade, and shelter, for soil protection and stabilization, for reclamation of derelict sites, and for their nature conservation and sporting values.

Generally, if trees are grown for a product other than wood, the enterprise becomes associated with agriculture or horticulture, though there is no real distinction between growing trees for timber or other products. In spite of this, in many temperate regions, a sharp division has been made in the last two centuries for management purposes, greatly hindering good integrated land husbandry. In recent years there has, however, been a revival of agroforestry practices, which now take various modern forms. Forestry is increasingly called upon to address several of the above aspects and 'multiple-use' management is becoming the normal practice.

The main types of plantations established for purposes other than timber are discussed.

Shelter

Using trees to reduce wind speed locally is important in arable and livestock farming and for fruit orchards. Historically, the main function of early hedgerows was to create enclosures for domestic animals, and during the nineteenth century they acquired many other roles including the provision of shelter, erosion control, and wood production. With the widespread mechanization of agriculture, massive destruction of hedgerows occurred during the twentieth century, particularly between 1950 and 1980 (Bazin 1993; Bazin and Schmutz 1994). Much research on the value of trees for shelter started on the Great Plains of the United States, the Canadian prairies, and Russian steppes. Denmark and Italy developed important shelterbelt planting programmes at the turn of the nineteenth century. In the 1980s, a revival of interest in shelterbelts occurred worldwide, with the creation of a working group on the subject within IUFRO and several symposia on 'shelterbelts and agroforestry' (Anon. 1993). The main beneficial effects of shelter are in increasing crop yields and reducing or preventing soil erosion.

Effects of tree shelterbelts on wind

Windbreaks or shelterbelts have the effect of reducing soil erosion by decreasing surface wind shear stresses, trapping moving soil, and slowing soil drying. Windbreak design is important and porosity, height, length, and orientation are the main considerations. When surfaces are highly erodible and wind speeds are above the threshold velocity necessary to initiate particle movement, the erosion rate is proportional to the cube of the wind speed. Thus, even modest reductions in wind speed can cause major reductions in erosion. Caborn (1965), Rosenberg (1974), Hagen (1976), Litvina and Takle (1993), and Olesen (1993) discuss, in detail, how shelterbelts affect wind.

  The taller a shelterbelt, the greater is the distance which is protected downwind, and to some extent, upwind as well. In general, a dense barrier provides some sheltering effect to a distance of 10 to 15 shelterbelt heights downwind. However, by increasing porosity to about 50 per cent, the downwind influence, though slightly reduced in degree, can be extended over a larger distance of 20 to 25 heights (Fig. 16.1), and also the somewhat greater wind penetration reduces turbulence with important consequences for lodging of crops and incidence of snowdrifts. The longer a windbreak the more constant its influence; if it is too short or has gaps, jetting effects may increase wind speed at the ends and near gaps. A minimum useful length is generally considered to be 12 times the final shelterbelt height. Ideally shelterbelt density should increase with height in proportion to the logarithmic increase in a wind-speed profile with height. Belts of trees are particularly effective in this respect, having dense crowns but, if pruned, can be relatively open nearer the ground. When shelterbelts are numerous enough to increase the overall surface roughness of the landscape the open field wind speeds can also be

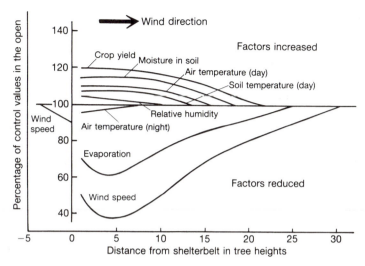

Fig. 16.1 Some effects of shelter on production and microclimate, based on research in Denmark. The curves do not show actual values but illustrate general trends. (From Olesen 1979, based on Marshall 1967.)

significantly reduced. A reduction in velocity of about 50 per cent compared with the open sea has been found in Denmark, in areas with many shelterbelts, but only 20 per cent in open, treeless country (Olesen 1979, 1993).

Effects of shelter on production

Apart from reducing dust storms and soil loss, shelter can have beneficial influences on crop and livestock production by modifying the microclimate to prevent excessive chilling, desiccation, or mechanical injury.

The beneficial and other effects on crop production have been discussed in detail by Rosenberg (1974), Guyot *et al.* (1986), and Fu (1993). Most are attributable to a reduction in the rate of evapotranspiration from the leaves of plants, caused by lower wind speeds and turbulence, and a consequent higher humidity in shelter. Plant growth and yields are usually greater in sheltered areas indicating that the net assimilation of carbon dioxide must be increased. This is ascribed partly to the longer period of photosynthetic activity in sheltered plants, owing to the delay or avoidance of wilting, and partly to the lower nocturnal temperatures found in sheltered zones (Fig. 16.1) which results in a reduction in respiration and hence an increase in net assimilation. By contrast, there is a greater water use in exposed areas because there is less resistance to the transport of water vapour. The rate of soil drying is increased, but over long enough periods, soils in both sheltered and exposed sites will reach the same degree of dryness. It is possible that the reduced stomatal resistance in sheltered zones may lead to greater evapotranspiration at times because daytime temperatures are higher and the plants larger.

The generally moister conditions in sheltered areas lead to more rapid seed germination, more vigorous vegetative growth, higher yields, and a reduction in mechanical injury, such as sand blasting, to plants. Shelter is also said to result in higher protein contents in wheat and, possibly, in greater oil and sugar yields from sunflower seeds and sugar beet respectively (Labaznikov 1982; Fu 1993).

Depending upon the climate, increases in crop yields of 20 per cent to as much as 150 per cent have been quoted in continental sheltered zones. However, growth and yield within one to two shelterbelt heights of a belt can be reduced owing to competition for light, water, and nutrients, and the land occupied by the belt itself is unavailable for crop production. The net effect on yields may therefore sometimes be quite small (Skidmore 1976), unless the timber production is also included.

In some maritime climates such as those of Britain and Ireland where humidity is generally high and where serious droughts seldom occur, excessive shelter can be harmful, slowing the drying of grain and increasing damage from disease in damp years. In upland areas, shelter can encourage earlier growth of grass in spring, providing a valuable 'early bite' for grazing animals (Figs 16.2 and 16.3). If there is enough protection, better species of grasses can be encouraged to grow.

Shelterbelts can also benefit livestock by providing shelter from strong winds. In western Europe, sheep are often lambing or in the early stages of lactation when grass growth begins on the hills, thus the improved growth induced by shelter is of

Fig. 16.2 A 2-storeyed New Zealand 'timberbelt' with high pruned *Pinus radiata* being grown for timber and unpruned cypress below for shelter.

Fig. 16.3   Sheep grazing in a *Pinus pinea* forest near Fréjus, Var, France. (Photo: M. Etienne, INRA.)

value as is the reduction in lamb losses in cold springs. Cattle are more sensitive to snow and cold winds and will usually seek shelter. By providing shelter, it is sometimes possible to keep cattle where previously only sheep could be stocked or to keep animals in winter on ground where it was not hitherto possible.

Shelterbelt design

After the massive destruction of hedgerows between 1950 and 1980, a new trend has appeared with the increase in environmental consciousness world-wide, particularly throughout Europe. It has now become clear that lines of trees can provide a large range of utilities, including fencing, shelter, reduction of wind and water erosion, wood production, landscape enhancement, and a source of biodiversity (Bazin 1993). Hedgerows are now being designed to fulfil several of these roles.

In New Zealand, a new concept has appeared, where some of the trees in shelterbelts are pruned in order to produce high quality timber. Different numbers may be pruned, depending on the main objective of the belt—windbreak or wood production. These 'timberbelts' usually consist of two strata: high pruned fast-growing *Pinus radiata*, underplanted with a slower growing species (Fig. 16.2) (Tombleson and Inglis 1988). The yield of well-managed timberbelts can be very high, with timber quality equal to that of plantations (Auclair *et al.* 1991; Tombleson 1993).

Several different systems are being developed in Europe. The Danish shelterbelt plantation scheme favours large 3- to 6-row belts, similar to some Australian and north American practices (Jørgensen 1993), whereas in France there is a trend towards one-row, multiple-use belts with two to three strata: a shrub layer for low shelter, a more wind-permeable coppice layer for fuelwood production, and some high quality timber-producing trees (IDF 1981; Schmutz 1994). The landscape value is often considered very beneficial in Britain.

Agroforestry

Agroforestry—defined as an intimate mixture of trees with farm crops and/or animals on the same piece of land—was the common practice in Europe in ancient times, and remains the main management system in a great many parts of the world (Nair 1993). In developed countries, the widespread use of heavy machinery and chemicals in agriculture, the land tenure with relatively large properties, and the considerable decrease in forest cover due to over-exploitation up to the end of the eighteenth century, have led to a separation between agriculture and forestry, and relegated agroforestry to marginal areas. However, in recent years the evolution of the European Union's Common Agricultural Policy towards more integrated land management and environmentally sound practices has prompted European researchers and land managers to develop modern forms of agroforestry (Guitton 1994; Auclair 1995*a*; Sheldrick and Auclair 1995).

Traditional agroforestry

Three main types of agroforestry are traditional in Europe, and still remain in some areas considered as marginal: hedgerows which have been discussed above, forest grazing, and tree/crop mixtures.

Forest grazing

Grazing sheep, cattle, or pigs (Fig. 16.4) in the forest was a traditional practice in the past, and can still be seen in some areas or at some periods in the year. In uplands, many farmers own small woodlots which provide shelter during cold spells in early spring or late autumn. These woodlands are sometimes used at lambing time, or to let animals into during mild periods in winter. In southern Europe where summer droughts are common, the shade inside forests can delay desiccation of the grass in late spring and early summer, providing an additional fodder resource at a time when the open pastures have dried out. Careful livestock management is necessary in silvopastoral systems, since too high a stocking over time or in animal numbers can lead to bark stripping, trampling and soil compaction, and browsing of young trees (Fig. 16.3).

Fig. 16.4 The acorns of *Quercus suber* and *Q. ilex* are an important seasonal source of food for pigs in Portugal and Spain. This food results in valuable hams with an excellent taste!

In *Larix* communities of the southern French Alps, the open crown of *Larix decidua* favours the growth of grass which is an important source of fodder for sheep, cattle, or horses at intermediate periods before and after grazing the Alpine pastures. In the Swiss Jura mountains, open communities of *Abies alba* and *Picea abies* are grazed by cattle. Many montane rangelands consist of a mosaic of vegetation containing forested areas, with pioneer trees such as *Pinus sylvestris* (Étienne 1995).

Large numbers of trees are used for the fodder value of their leaves or fruits. The best known examples are *Quercus* spp., *Castanea sativa*, and *Fagus sylvatica*, whose fruits can be an important resource for domestic animals. Shrubby or woody ground vegetation, and a number of multipurpose trees are often used in southern regions (Talamucci 1989). There are no temperate trees as multipurpose and highly valued as the Mediterranean carob (*Ceratonia siliqua*). The pods of this leguminous species have a very high sugar content (40–59 per cent by weight) and are of great value as a supplement to animal feeds. The seeds are sometimes used for such diverse purposes as in the manufacture of chocolate and for making colloidal protection agents for offshore drilling (Winer 1980). The tree is also used for fuelwood and it thrives on land too dry, infertile, or rocky to support most other crops.

Agrisilviculture

Trees have generally been eliminated from arable fields with the development of mechanization, but some relatively marginal agrisilvicultural systems still persist in

Europe. Intercropping fruit orchards for vegetable production can be considered as a purely agricultural system if no timber is produced. However, many walnut (*Juglans regia*) orchards in France which are managed both for fruit and timber (p. 242) are very often intercropped with maize during the early years, and *Populus* plantations are sometimes intercropped for several years. Large cultivation-free areas must be kept around the trees in order to avoid damage from machinery, and careful attention must be given to the herbicides applied to crops in order to avoid damage to the trees.

Modern trends in agroforestry

Today, in conjunction with the Common Agricultural Policy, traditional European agroforestry is being rejuvenated and new, more intensive systems are being developed (Guitton 1994; Sheldrick and Auclair 1995).

Silvopastoral systems

Modern, scientific interest in temperate zone silvopastoral agroforestry started in New Zealand in the 1960s and has expanded rapidly, based on the production of *Pinus radiata* (Knowles 1991). In Europe, two main silvopastoral systems are being developed: modern forest grazing and timber-producing trees on pasture.

In order to reduce the establishment and management costs, modern 'dynamic' silviculture requires wide spacings, which are made possible by the use of high quality genetic stock. To control understorey vegetation, improve access for forest workers, and reduce the hot dry season fire hazard (Chapter 13), domestic animals can be introduced to the forest as soon as the trees become large enough to withstand grazing. Livestock can thus be considered as a silvicultural tool, but proper integrated silvopastoral management is necessary both to avoid possible damage to the tree crop and to satisfy the livestock farmer by providing adequate fodder. Oversowing with adapted species, such as fodder grasses, clover, or *Lotus uliginosus*, eliminating non-palatable and thorny understorey species, and adding fertilizer often prove useful (Rapey *et al.* 1994).

To provide for land-use diversification and help reduce agricultural overproduction, one solution, which has been largely developed in New Zealand, consists of planting widely spaced timber-producing trees on pasture. Research is under way to adapt this technique to European conditions, and to study the socioeconomic impacts. The first results using the following techniques seem very promising (Rapey 1994; Auclair 1995*b*):

♦ selection of high quality timber-producing species, adapted to the site;

♦ 50 to 400 seedlings planted per hectare produced using good nursery practices;

♦ adequate soil preparation;

♦ local control of vegetation around the trees for up to 5 years;

♦ individual tree protection;

◆ individual tree formative pruning;

◆ stocking of livestock adapted to the forage supply.

Silvoarable systems or agrisilviculture

China offers a range of traditional forms of agroforestry with some highly impressive modern agrisilvicultural systems based on *Paulownia tomentosa* underplanted with wheat, beans, and medicinal plants (Newman 1994). Although the earliest modern British trial with *Populus* spp. underplanted with cereals collapsed under economic pressures, subsequent trials in Britain, France, and Italy with the latest genetically improved *Populus* clones are encouraging (Beaton *et al.* 1995).

Systems similar to these, consisting of high-quality timber trees underplanted with arable crops, are being studied but are still too young to produce definite results (Dupraz 1994; Incoll *et al.* 1995). The suggested management is derived both from traditional walnut agrisilvicultural systems and silvopastoral experiments. They are all similar, with tree spacings of approximately 13×4 m, depending on the machinery to be used for sowing and spraying. Individual tree shelters are advised, to protect them from mechanical damage and chemical sprays. A clean strip of 2 m should be kept either side of the trees.

Amenity forestry

Urban forestry and community woodlands

Tree planting, to provide woodland amenity or to soften the urban landscape, is now commonplace. In poorly wooded countries like Britain and The Netherlands, many new urban woodlands managed for the benefit of, and often by, the local community have been established in the last 25 years. They call for a different approach to tree planting, and advice to meet the special needs is now available, for example Hodge (1995). Although the principles of silviculture to obtain good tree growth are the same as for any other site, the objects are very different: timber production has little importance while attractiveness, resilience, and naturalness of the woodland are critical to success. Above all a community woodland is a place where people should enjoy going.

Silvicultural priorities can be listed quite simply:

(1) planting species which will survive well and grow in the locality—often this means a tolerance of poor soils or artificial substrates since much urban forestry is on derelict land;

(2) combining mixtures of species of trees and shrubs, usually in groups, and at irregular spacings;

(3) sufficient attention to weed control and protection to ensure tree survival and health;

(4) above all, plenty of provision for open space—some of the best community woodlands have 25 to 30 per cent open ground.

As a new community woodland becomes established, the forester works with nature developing the best and often not worrying about poor areas or even failures. There is no economic demand to stock the ground fully, and the balance of success and failure can be turned to advantage to lend more naturalness. It is very different from the production orientated timber plantation, but it is still a product, and a much valued one, of employing the forester's skill in tree planting and woodland creation.

Creating new native woodlands

Much plantation forestry has used exotic trees with great success. For timber plantations such a choice has often proved the most productive as species are matched with sites, endemic pests and diseases are absent, and the tree flourishes in its new environment. Spectacular successes include *Pinus radiata* in many southern hemisphere countries, *Eucalyptus globulus* in Spain, Portugal, and Turkey, and many western North American conifers in the moist temperate parts of Europe. However, today, tree planting aims to satisfy wider purposes than wood production, including recovery of impoverished native forest and woodland, often destroyed in previous centuries by agricultural expansion.

Creating new native woodland involves much more than planting native trees. To recreate the woodland types formerly characteristic of a locality successfully requires an understanding of regional ecology and a robust vegetation classification on which scientifically sound silvicultural decisions can be taken. This has been attempted in Britain with the provision of general guidance by Rodwell and Patterson (1994) and specifically to encourage expansion of native *Pinus sylvestris* in the Scottish Highlands.

Designing new native woodlands has to be location specific to fit what soils and landscape determine. Appropriate natural colonization can often be preferable to planting, and when planting, consideration may need to be given to using local genotypes if these are known and available. In addition, a native woodland is more than just its trees. The practice of introducing field layer plants, particularly woodland specialists, though still in its infancy, may be increasingly sought. Natural colonization of a truly woodland field layer is very slow, even where the new wood adjoins an old one. Deliberate introduction will speed the process, with initial indications suggesting the use of container-grown plants as superior to direct sowing.

Field sports

For many owners the hunting, stalking, and shooting of wildlife they derive from their woodlands can be an important return. On some estates and farms much planting is done principally for this purpose and any yield of timber or benefit of shelter is only a secondary consideration.

Other products from trees and forests

Exudates

Tree exudates which are of value to man include latex, resins, gums, and sugary saps from which products such as maple syrup and rosin are made.

Gums and resins

In France, the resin industry dates back to the 1700s and was stimulated by the need for land reclamation to prevent severe damage from drifting sand, mostly in the Landes. Planting started in 1803 using primarily *Pinus pinaster* which was subsequently tapped for resin, as well as used for timber. Significant amounts of tapping are also done in Spain, Portugal, and Greece, with *P. pinaster, P. halepensis*, and *P. nigra*, and in Russia with *P. sylvestris*.

Modern uses of resins are in the manufacture of paint, varnish, and lacquers, as size in paper making to provide lustre and weight and to hinder the absorption of ink and moisture, and in making soap. They are also used in pharmaceutical and chemical industries.

Aromatic oils

Since early historical times, aromatic oils found in parts of various plants have attracted man. They are known as volatile or essential oils, and have uses in medicine, as inhalants, embrocations, soaps, and antiseptics, or in industry. In Portugal and Spain oil is obtained from *Eucalyptus globulus* foliage grown in plantations.

Sap

Maple syrup is one of North America's oldest commodities. Its production has always been a farm enterprise and though production has declined steadily since the peak in 1860, it is still very profitable. Management of *Acer saccharum* has been discussed by Smith and Gibbs (1970) and Lancaster *et al.* (1974). *Betula* is tapped in Ukraine forests which are managed primarily for timber production (Sendak 1978). The main sugars in *Betula* sap are glucose and fructose with some sucrose, whereas there is only sucrose in maple sap.

Nectar

Bee-keeping depends on plentiful flowering within the vicinity for success. Trees are able to augment wild flowers and, in some instances, become a primary source of nectar, for example *Acer, Castanea,* and *Robinia*.

In some instances the selection and breeding has included flowering habit because of the nectar potential; one example is false acacia (*Robinia pseudoacacia*) which is widely grown in Hungary for timber. Varieties have been bred which show both improved timber production and a prolongation of the flowering period to increase nectar yields for bee-keepers (Keresztesi 1983).

Ornamental foliage and Christmas trees

In many temperate countries there is a long tradition of using decorative foliage. Methods of production of ornamental foliage and Christmas trees (mainly from young *Picea abies* and various species of *Abies*) range from being lucrative supplements to normal forest practice to highly specialized enterprises in their own right. In some parts of Europe, it is possible to dispose profitably of the lower branches of trees such as *Chamaecyparis lawsoniana*, *Tsuga heterophylla*, and *Thuja plicata* to florists and to sell the tops of thinned *Picea abies* as Christmas trees.

Denmark produces large quantities of foliage of *Abies procera* and *A. nordmanniana* from strains with specially selected characteristics such as glaucousness and greenness (Barner *et al.* 1980). In France and Britain, *Eucalyptus* crops are grown specifically for the attractive sprays of glaucous or bluish juvenile foliage produced by blue gums and white gums. Christmas trees are sometimes interplanted in normal forest stands to provide early financial returns to help offset establishment costs, or planted pure on small, easily accessible areas. When stands are grown mainly for greenery or Christmas trees, care must obviously be taken to discourage defoliating insects, and the removal of foliage represents a considerable export of nutrients from the site, with little recycling of litter. To correct this drain, fertilizers are usually applied to maintain yields (Holstener-Jørgensen 1977).

Bark

The bark of a large number of species is removed prior to using the timber and is pulverized and sold for mulch or as a soil conditioner, and for the manufacture of some kinds of building boards. These can be lucrative means of disposing of what might otherwise be a waste product (Aaron 1982). The tan bark industry was, at one time, important in many temperate countries. In much of Europe, it was based on oak (*Quercus robur*).

Cork is an important bark product from *Quercus suber* in Mediterranean regions (Fig. 16.5). Its cultivation, uses, and properties have been discussed in detail by de Oliveira and de Oliveira (1991). Cork is the outer bark of the tree, and is stripped from the stem and sometimes from the larger lower branches when the trees are 20 to 30 years old and thereafter every 9 or 10 years during their productive life of about 100 years. Cork oak (*Quercus suber*) is often grown in conjunction with raising pigs which thrive on the acorns. It is also planted as a shade tree

Fig. 16.5 *Quercus suber* being grown for cork production in Portugal.

in vineyards. Research on silvicultural management and genetic improvement of cork-oak stands is being carried out in southern Europe (Anon. 1992; Pereira 1995) though many stands are suffering from dieback and decline.

The main properties of cork are buoyancy, compressibility, resilience to moisture and liquid penetration, high frictional quality, low thermal conductivity, ability to absorb vibration, and stability. Although synthetic materials such as polystyrene and glass fibre have replaced cork for many uses, demand for cork exceeds supply and increasingly cork composites are used, mainly for their 'natural' appearance, thus improving its utilization.

Food

Fruit, nuts, and fodder

The importance of tree fruits, seeds, and leaves in human and animal diets is enormous in some parts of the world. Few, however, are of major importance in temperate regions, except where cultivated varieties of trees are specially planted in orchards or as single trees in gardens or elsewhere.

Among valuable timber species, the common walnut, *Juglans regia*, is an important temperate nut-bearing tree, which has long been grown in Italy and in France (Périgord and Dauphiné), where numerous varieties have been selected for their fruit production. In Europe, walnuts are cultivated either as individual trees in farmyards or in orchards. In many walnut orchards, production of both fruit and timber is sought: the trees are pruned in order to produce a 2 to 4 m knot-free bole which can yield a very high timber value at the end of the fruit-producing cycle. The black walnut (*Juglans nigra*) is an important forest timber tree in Hungary and in the United States, but the nuts are of less importance than the former species because of the thick, hard shells.

Several species of chestnut are important sources of food. *Castanea sativa*, in particular, has been cultivated for this purpose for centuries in southern Europe and was one of the staple foods in Roman times. The nuts were used as a source of flour and also for feeding animals. They are rich in starch and sugar but contain relatively little oil or fat and are therefore nutritionally akin to cereals. The famous 'marrons glacés' are produced in French and Italian chestnut orchards, from grafted trees selected for fruit size and quality. Although traditional chestnut cultivation is declining in many regions, chestnut gathering is a common practice in autumn in southern Europe.

Pine seeds, specially those of the southern European stone pine, *Pinus pinea*, are also used for food or as dessert nuts in various parts of the world.

Fungi

The cultivation of edible fungi has little influence on normal forest management since many such fungi are mycorrhizal associates of forest trees. Others are saprophytes on rotting wood and a few (e.g. *Armillaria* spp.) are sometimes pathogenic. One of the best known and most valued of fungi is the truffle (*Tuber melanosporum*), a mycorrhizal species on *Quercus*, *Corylus avellana*, *Carpinus betulus,* and *Castanea sativa*. The value of truffles is so great in France and Italy that techniques have been developed for inoculating tree seedlings in the nursery specifically for truffle production (Gregori and Tocci 1990), and truffles are now produced in specific orchards.

Multiple-use forest management

Although forest plantations are very often managed with wood production as the main objective, the present general environmental concerns strongly influence management. Forests are often considered by the public as 'natural' systems—even artificial plantations as they grow mature. There are generally many demands on forests, arising from the large number and variety of interests. Forests are expected to address many of the aspects mentioned in this chapter, often simultaneously:

shelter for crops or for animals, amenity, landscape, biodiversity, hiking, biking, shooting, and other products such as fungi or berries. Such 'multiple-use' forest management is becoming the normal practice. It requires not only knowledge of silviculture and economics, but also sociological skills in order to take into account all the demands and to deal in a satisfactory manner with the variety of interests (Bland and Auclair 1996). Further information can be found in Insley (1988), Hibberd (1989), and Bazin (1992).

As Grayson (1993) has pointed out in his review of forest policy in western Europe, the growing point of forestry debate continues to be the scale and kind of adjustment desired to meet the demands for more beautiful, more varied, and more wildlife-rich forests. Compromises have to be found between those whose incomes will be affected by the changes and those whose interest is broadly in the achievement of some aesthetic, ethical, or other non-monetary objective. Planting trees will continue to play an important role to provide these many benefits.

References

Aaron, J.R. (1982). *Conifer bark: its properties and uses.* Forestry Commission Forest Record, 110. HMSO, London.

Adams, S.N., Dickson, D.A., and Cornforth, I.S. (1971). Some effects of soil water tables on the growth of Sitka spruce in Northern Ireland. *Forestry*, **45**, 129–33.

Agestam, E. (1985). [*A growth simulator for mixed stands of pine, spruce and birch in Sweden.*] Department of Forest Yield Reseach, Swedish University of Agriculture Sciences, Garpenberg, Report No. 15.

Agrios, G.N. (1978). *Plant pathology.* New York.

Agyeman, J. (1996). *Involving communities in forestry...through community participation.* Forestry Practice Guide, 10. Forestry Commission, Edinburgh.

Alder, D. (1980). *Forest volume estimation and yield prediction.* FAO Forestry Paper, 22/2.

Aldhous, J.R. (1989). Standards for assessing plants for forestry in the United Kingdom. *Forestry*, **62** (suppl.), 13–19.

Aldhous, J.R. and Mason, W.L. (ed.) (1994). *Nursery practice.* Forestry Commission Bulletin, 111. HMSO, London.

Alexandrian, D. and Gouiran, M. (1990). Les causes d'incendie: levons le voile. *Revue Forestière Française*, **42**, No. special, 33–41.

Anderson, M.L. (1961). *The selection of tree species* (2nd edn). Oliver & Boyd, London.

Andersson, S.-O. (1963). [*Yield tables for plantations of Scots pine in north Sweden*]. Meddelanden från Statens skogsforsknings institut, 44. Stockholm.

Anon. (1973). [The Göttingen Conference on forest and storm damage, February 1973.] *Forstarchiv*, **44**, 41–75.

Anon. (1976). *Water balance of the headwater catchments of the Wye and Severn.* Institute of Hydrology Report, 33. NERC, Wallingford, UK.

Anon. (1992). *EEC Symposium on cork biology.* Scientia Gerundensis, 18. Universitat Girona, Spain.

Anon. (1993). *Windbreaks and agroforestry.* Hedeselskabet, Viborg, Denmark.

Anon. (1994). *Skogsvårdslagen.* Swedish Board of Forestry.

Armstrong, W., Booth, T.C., Priestley, P., and Read D.J. (1976). The relationship between soil aeration, stability and growth of Sitka spruce. *Journal of Applied Ecology*, **13**, 585–91.

Arnold, J.E.M. (1983). Economic considerations in agroforestry projects. *Agroforestry Systems*, **1**, 399–411.

Arnold, J.E.M. (1991). *Community forestry: ten years in review*. Community Forestry Note, 7. FAO, Rome.

Arnold, J.E.M. and Stewart, W.C. (1991). *Common property resource management in India.*, Tropical Forestry Papers, 24. Oxford Forestry Institute, Oxford.

Ascher, W. and Healy, R.G. (1990). *Natural resource policy making in developing countries*. Duke University Press, Durham, NC.

Assmann, E. (1955). Die Bedentung des 'erweiterten Eichhorn'schen Gesetzes' fur die Konstruktion von Fichten–Ertragstafeln. *Forstwissenschaftliches Centralblatt*, **74**, 321–30.

Assmann, E. (1970). *The principles of forest yield study*. Pergamon Press, Oxford.

Assmann, E. and Franz, F. (1965) Vorläufige Fichten-Retragstafel fur Bayern. *Forstwiss*, **84**, 13–43.

Auclair, D. (1995a). Agroforesterie: Intégration et entretien de la production ligneuse dans l'exploitation agricole. In *Sixteenth COLUMA conference, International meeting on weed control*. Annales ANPP, Vol. **I/III**, 135–44.

Auclair, D. (1995b). Alternative agricultural land-use with fast-growing trees: Technical research and development of a biologically-based economic modelling system for Europe. In *Agroforestry and land-use in industrialized nations* (ed. E. Welte, I. Szabolcs, and R.F. Hüttl), pp. 251–73. Goltze–Druck, Göttingen.

Auclair, D. and Bouvarel, L. (1992). Intensive or extensive cultivation of short rotation hybrid poplar coppice on forest land. *Bioresource Technology*, **42**, 53–9.

Auclair, D. and Cabanettes, A. (1987). Short rotation poplar coppice production compared to traditional coppice. In *Biomass for energy and industry* (ed. G. Grassi, B. Delmon, J.F. Molle, and H. Zibetta), pp. 551–5. Elsevier, London.

Auclair, D., Bédéneau, M., Cabanettes, A., and Pontailler, J.Y. (1988). Quelques aspects du fonctionnement des taillis. In *Les phénomènes de réitération chez les végétaux, 1ère réunion du Groupe d'Étude de l'Arbre*, Grenoble. CEA, Centre d'Études Nucléaires de Grenoble, pp. 26–42.

Auclair, D., Tombleson, J., and Milne, P. (1991). *A growth model for Pinus radiata timberbelts*. N.Z. Ministry of Forestry, Forest Research Institute, Rotorua.

Baker, F.S. (1934). *Theory and practice of silviculture*. McGraw–Hill, London.

Baker, H.G. (1974). The evolution of weeds. *Annual Review of Ecology and Systematics*, **5**, 1–24.

Ballard, R. (1986). Fertilization of plantations. In *Nutrition of plantation forests* (ed. G.D. Bowen and E.K.S. Nambiar), pp. 327–60. Academic Press, London.

Barbier, E.B. (1987). The concept of sustainable economic development. *Environmental Conservation*, **14**, 101–10.

Barner, H., Rouland, H., and Qvortrup, S.A. (1980). [*Abies procera* seed supply and choice of provenance.] *Dansk Skovforenings Tidsskrift*, **65**, 263–95.

Barnes, D.F. and Olivares, J. (1988). *Sustainable resource management in agriculture and rural development projects: a review of bank policies, procedures and results.* World Bank, Washington DC, USA.

Barnes, R.D. and Mullin, L.J. (1989). The multiple population breeding strategy in Zimbabwe—five year results. In *Breeding tropical trees* (ed. G.L. Gibson, A.R. Griffin, and A.C. Matheson), pp. 148 58. Oxford Forestry Institute and Winrock International.

Barthod, C., Buffet, M., and Sarrauste de Menthière M. (1990). Le contexte technico économique de l'emploi des herbicides en forêt. In *Journées internationales d'études sur la lutte contre les mauvaises herbes.* Annales A.N.P.P. No. 3, 845–60. Paris.

Bastien, J.C. and Demarcq, P. (1994). *Choix d'espèces pour les reboisements d'altitude dans le Massif Central.* Office National des Forêts, Bulletin Technique, **27**.

Baule, H. and Fricke, C. (1970). *The fertilizer treatment of forest trees.* BLV Verlagsgesellschaft mbH, Munich, Germany.

Bazin, P. (1992). *Boiser une terre agricole.* IDF, Paris.

Bazin, P. (1993). Situation of hedgerows and shelterbelts in the EEC. In *Windbreaks and agroforestry*, pp. 129–132. Hedeselskabet, Viborg, Denmark.

Bazin, P. and Schmutz, T. (1994). La mise en place de nos bocages en Europe et leur déclin. *Revue Forestière Française*, **46**, 115–18.

Beaton, A., Burgess, P., Stephens, W., Incoll, L.D., Corry, D.T, and Evans, R.J. (1995). Silvoarable trial with poplar. *Agroforestry Forum*, **6**, 14.

Bédéneau, M. and Auclair, D. (1989). A comparison of coppice and single-stem root distribution using spiral trenches. *Acta Œcologica, Œcologia Applicata*, **10**, 293–220.

Behan, R.W. (1990). Multiresource forest management: a paradigmatic challenge to professional forestry. *Journal of Forestry*, **88**, 12–18.

Belyi, G.D. (1975). [Stand density and its regulation in the control of *Fomes annosus.*] *Leso Vodstvo i agrolesmelioratsiya*, **40**, 28–35.

Bergez, J.É, Auclair, D., and Bouvarel, L. (1989). First year growth of hybrid poplar shoots from cutting or coppice origin. *Forest Science*, **35**, 1105–13.

Bevan, D. (1984). Coping with infestations. *Quarterly Journal of Forestry*, **78**, 36–40.

Bevege, D.I. (1984). Wood yield and quality in relation to tree nutrition. In *Nutrition of plantation forests* (ed. G.D. Bowen and E.K.S. Nambiar), pp. 293–326. Academic Press, London.

Bingham, C.W. (1985). Rationale for intensive forestry investment: a 1980s view. In *Investment in forestry* (ed. R.A. Sedjo), pp. 21–31. Westview Press, Boulder, CO, USA.

Binns, W.O. (1962). Some aspects of peat as a substrate for tree growth. *Irish Forestry*, **19**, 32–55.

Binns, W.O. (1975). *Fertilizers in the forest: a guide to materials.* Forestry Commission Leaflet, 63. HMSO, London.

Binns, W.O. and Crowther, R.E. (1983). Land reclamation for trees and woods. *Reclamation*, **83**, 23–8.

Binns, W.O., Mayhead, G.J., and Mackenzie, J.M. (1980). *Nutrient deficiencies of conifers in British forests.* Forestry Commission Leaflet, 76. HMSO, London.

Bland, F. and Auclair, D. (1996). Silvopastoral aspects of Mediterranean forest management. In *Temperate and mediterranean silvopastoral systems of western Europe* (ed. M. Étienne), pp. 125–42. Colloques de l'INRA, Versailles, France.

Blatner, K.A. and Greene, J.L. (1989). Woodland owner attitudes toward timber production and management. *Resource management and optimization*, **6**, 205–23.

Boggie, R. (1972). Effect of water-table height on root development of *Pinus contorta* on deep peat in Scotland. *Oikos*, **23**, 304–12.

Boggie, R. and Miller, H.G. (1976). Growth of *Pinus contorta* at different water-table levels in deep blanket peat. *Forestry*, **49**, 123–31.

Boland, D.J. (ed.). (1989). *Trees for the tropics.* ACIAR, Canberra, Australia.

Bollen, W.B., Chen, C.S., *et al.* (1967). *Influence of red alder on fertility of a forest soil. Microbial and chemical effects.* Research Bulletin Oregon Forest Research Laboratory, 12.

Bonneau, M. (1978). L'analyse foliaire. *La Forêt Privée*, January, pp. 31–5.

Bonneau, M. (1986). Fertilisation à la plantation. *Revue Forestière Française*, **38**, 293–300.

Booth, T.H. (1995). *Where will it grow? How well will it grow?* Australian Centre for International Agricultural Research, RN 16.

Bornebusch, C.H. (1937). [Report on the incidence of storm damage in the spruce thinning plots in Hastrup Plantation.] *Forstlige Forsøgsvaesen i Danmark*, **14**, 161–72.

Boudru, M. (1986). *Forêt et sylviculture 1. Sylviculture appliquée.* Les Presses Agronomiques de Gembloux.

Boudru, M. (1989). *Forêt et sylviculture 2. Traitement des forêts.* Les Presses Agronomiques de Gembloux.

Bowen, G.D. (1984). Tree roots and the use of soil nutrients. In *Nutrition of plantation forests* (ed. G.D. Bowen, and E.K.S. Nambiar), pp. 147–79. Academic Press, London.

Braastad, H. (1974). [Diameter increment functions for *Picea abies*]. *Meddelelser fra Norsk Institutt for Skogforskning*, **31** (1), 1–74.

Braastad, H. (1978). [Thinning intensity and frequency of damage. Report on snow and wind damage in research plot 918.] *Norsk Skogbruk*, **24**, 20.

Braastad, H. (1980). [Growth model computer program for *Pinus sylvestris*]. *Meddelelser fra Norsk Institutt for Skogforskning*, **35** (5), 265–359.

Bradley, R.T. (1963). Thinning as an instrument of forest management. *Forestry*, **36**, 181–94.

Bradley, R.T., Christie, J.M., and Johnston D.R. (1966). *Forest management tables*. Forestry Commission Booklet, 16. HMSO, London.

Bradshaw, A.D. and Chadwick, M.J. (1980). *The restoration of land*. Blackwell, Oxford.

Brazier, J.D. and Mobbs, I.D. (1993). The influence of planting distance on structural wood yields of unthinned Sitka spruce. *Forestry*, **66**, 333–52.

Brinar, M. (1972). [Investigation of the hereditary characters of a particular selected spruce.] *Gozderski Vestnik*, **30**, 37–45.

Brouard, N.R. (1967). *Damage by tropical cyclones to forest plantations with particular reference to Mauritius*. Government Printer, Mauritius.

Brünig, E.F. (1973). [Storm damage as a risk factor in wood production in the most important wood-producing regions of the earth.] *Forstarchiv*, **44**, 137–40. (trans. W. Linnard, Commonwealth Forestry Bureau, Oxford No. 4339).

Brunton, D.P. (1987). Financing small-scale rural manufacturing enterprises. In *Small-scale forest-based processing enterprises*, pp. 117–48. Forestry Paper 79. FAO, Rome.

Bryant, J.P., Chapin, F.S., and Klein, D.R. (1983). Carbon/nutrient balance of boreal plants in relation to vertebrate herbivory. *Oikos*, **40**, 357–68.

Bucknell, J. (1964). *Climatology—an introduction*. Macmillan, London.

Burcschel, P. and Huss, J. (1987). *Grundriss des waldbaus*. Verlag Paul Parey, Hamburg.

Burdett, A.N. (1978). Root form and mechanical stability in planted lodgepole pine in British Columbia. In *Proceedings of the root form of planted trees symposium*. British Columbia Ministry of Forests/Canadian Forestry service Joint Report, 8. (ed. E. van Eerdenan and J.M. Kinghorn), pp. 161–5.

Burdett, A.N. (1979). Juvenile instability in planted pines. *Irish Forestry*, **36**, 36–47.

Burdon, R.D. (1989). When is cloning on an operational scale appropriate? In *Breeding tropical trees* (ed. G.L. Gibson, A.R. Griffin, and A.C. Matheson), pp. 9–27. Oxford Forestry Institute and Winrock International.

Burgess, C.M., Britt, C.P., and Kingswell, G. (1996). The survival and early growth, in a farm woodland planting, of English oak from bare rooted and cell grown stock planted over five dates from September to May. *Aspects of Applied Biology*, **44**, 89–94.

Burke, M.J., Gusta, L.V., Quamme, H.A., Weiser, C.J., and Li, P.H. (1976). Freezing and injury to plants. *Annual Review of Plant Physiology*, **27**, 507–28.

Burke, W. (1967). Principles of drainage with special reference to peat. *Irish Forestry*, **24**, 1–7.

Burley, J. (1965). Genetic variation in *Picea sitchensis*. *Commonwealth Forestry Review*, **44**, 47–59.

Burschel, P. and Huss, J. (1987). *Grundriß des Waldbaus—Ein Leitfaden für Studium und Praxis*. Verlag Paul Parey, Hamburg.

Büsgen, M., Münch, E., and Thomson, T. (1929). *The structure and life of forest trees*. Chapman and Hall, London.

Butin, H. and Shigo, A.L. (1981). *Radial shakes and 'frostcracks' in living oak trees*. USDA Forest Service Research Paper NE–478.

Byron, R.N. and Waugh, G. (1988). Forestry and fisheries in the Asian–Pacific region: issues in natural resource management. *Asian–Pacific Economic Literature*, **2**, 46–80.

Cabanettes, A. (1987). Production de biomasse ligneuse dans les taillis traditionnels (France). In *Biomass for energy and industry* (ed. G. Grassi, B. Delmon, J.F. Molle, and H. Zibetta), pp. 556–61. Elsevier, London.

Caborn, J.M. (1965). *Shelterbelts and windbreaks*. Faber and Faber, London.

Cajander, A.K. (1913). Studien über die Moore Finlands. *Acta Forestalia Fennica*, **2**, 1–208.

Calabri, G. and Ciesla, W.M. (1992). *Global wildland fire statistics 1981–1990*. FAO Forest Resources Division, Rome.

Caldwell, L.K. (1990). *Between two worlds*. Cambridge University Press.

Callaham, R.Z. (1964). Provenance research: investigation of genetic diversity; associated with geography. *Unasylva*, **18**, 40–50.

Cameron, J.L. and Penna, I.W. (1988). *The wood and the trees*. Australian Conservation Foundation, Melbourne.

Cannell, M.G.R. (1980). Productivity of closely spaced young poplar on agricultural soils in Britain. *Forestry*, **53**, 1–21.

Cannell, M.G.R. (1988). The scientific background. In *Biomass forestry in Europe: a strategy for the future* (ed. F.C. Hummel, W. Palz, and G. Grassi), pp. 83–140. Elsevier, London.

Cannell, M.G.R. and Smith, R.I. (1980). Yields of minirotation closely spaced hardwood: temperate regions: review and appraisal. *Forest Science*, **26**, 415–28.

Carbonnier, C. (1954). [*Yield studies in planted spruce stands in southern Sweden*]. Meddelanden från Statens Skogsforskningsinstitut, 44, No. 5. Stockholm.

Carre, J. and Lemasle, J.M. (1987). Production of syngas by lignocellulosic biomass gasification. In *Biomass for energy and industry* (ed. G. Grassi, B. Delmon, J.F. Molle, and H. Zibetta), pp. 263–7. Elsevier, London.

Cassells, D.S. and Valentine, P.S. (1990). From conflict to consensus—towards a framework for community control of the public forests and wildlands. *Australian Forestry*, **51**, 47–56.

Chambers, R. and Leach, M. (1989). Trees as savings and security for the rural poor. *World Development*, **17**, 329–42.

Charles, P.J. and Chevin, H. (1977). Note sur le genre *Acantholyda*, et plus parti-culièrement sur *Acantholyda hieroglyphica*. *Revue Forestière Française*, **29**, 22–6.

Chavasse, C.G.R. (1979). The means to excellence through plantation establishment: the New Zealand experience. In *Forest Plantations: the shape of the future. Proceedings of Weyerhaeuser Science Symposium* Tacoma, Washington, 30 April–3 May 1978 (ed. D.D. Lloyd), pp. 119–37.

Cheliak, W.M. and Rogers, D.L. (1990). Integrating biotechnology into tree improvement programs. *Canadian Journal of Forest Research*, **20**, 452–63.

Cheney, N.P. (1971). *Fire protection of industrial plantations.* FO:SF/ZAM5, Technical Report 4. FAO, Rome.

Cheney, N.P. (1990). La gestion actuelle des incendies de forêts en Australie. *Revue Forestière Française*, **42**, N° sp., 368–74.

Chevrou, R., Delabraze, P., Malagnoux, M., and Velez, R. (ed.) (1995). *Forest fires in the Mediterranean region.* Options Méditerranéennes, série A, 25.

Christersson, L. (1994). The future of European agriculture: food, energy, paper and the environment. *Biomass and Bioenergy*, **6**, 141–4.

Christersson, L., Sennerby–Forsse, L., and Zsuffa, L. (1993). The role and significance of woody biomass plantations in Swedish agriculture. *Forestry Chronicle*, **69**, 687–93.

Christiansen, E. and Bakke, A. (1971). Feeding activity of the pine weevil *Hylobius abietis* L. during a hot period. *Norsk Entomologisk Tidsskrift*, **18**, 109–11.

Clutter, J.L., Fortson, J.C., Pienar, L.V., Brister, G.H., and Bailey, R.L. (1983). *Timber management: a quantitative approach.* Wiley, New York.

Cocklin, C.R. (1989). Methodological problems in evaluating sustainability. *Environmental Conservation*, **16**, 343–51.

Cole, D.W. (1986). Nutrient cycling in world forests. In *Forest site and productivity* (ed. S.P. Gessel), pp. 103–15. Martinus Nijhoff, Netherlands.

Cole, D.W. and Rapp, M. (1981). Elemental cycling in forest ecosystems. In *Dynamic properties of forest ecosystems* (ed. D.E. Reichle), pp. 449–76. Cambridge University Press.

Colin, F., Houllier, F., Joannes, H., and Haddaori A (1993). Modélisation du profil vertical des diamètres, angles et nombres de branches pour trois provenances d'Épicéa commun. *Silvae Genetica*, **42**, 4–5.

Commission Nationale du Peuplier (1995). État et perspectives de la populiculture. *Comptes Rendus de l'Académie d'Agriculture de France*, **81**, 1–272.

Conway, G. (1981). Man versus pests. In *Theoretical ecology* (2nd edn) (ed. R.M. May), pp. 356–86. Blackwell, Oxford.

Cook, C.C. and Grut, M. (1989). *Agroforestry in sub-Saharan Africa.* World Bank Technical Paper No. 112. World Bank, Washington, DC.

Cooke, A. (1983). The effects of fungi on food selection by *Lumbricus terrestris*. In *Earthworm ecology* (ed. J.E. Satchell), 365–73. Chapman & Hall, London.

Cooper, A.B. and Mutch, W.F.S. (1979). The management of red deer in plantations. In *Ecology of even-aged forest plantations* (ed. E.D. Ford, D.C. Malcolm, and J. Atterson), pp. 453–62. Institute of Terrestrial Ecology, Cambridge.

Costigan, P.A., Bradshaw, A.D., and Gemmell, R.P. (1982). The reclamation of acidic colliery spoil. III Problems associated with the use of high rates of limestone. *Journal of Applied Ecology*, **19**, 193–201.

Cotterill, P.P. (1986). Genetic gains expected from alternative breeding strategies including simple low cost options. *Silvae Genetica*, **35**, 212–23.

Countryside and Forestry Commissions (1991). *Forests for the community*. Countryside Commission, London.

Coutts, M.P. and Armstrong, W. (1976). The role of oxygen transport in the tolerance of trees to waterlogging. In *Tree physiology and yield improvement* (ed. M.G.R. Cannell and F.T. Last), pp. 361–85. Academic Press, London.

Coutts, M.P. and Grace, J. (ed.) (1995). *Wind and trees*. Selected papers from a conference held at Heriot–Watt University, Edinburgh, July 1993. Cambridge University Press.

Coutts, M.P. and Philipson, J.J. (1978*a*). Tolerance of tree roots to waterlogging: I. Survival of Sitka spruce and lodgepole pine. *New Phytologist*, **80**, 63–9.

Coutts, M.P. and Philipson, J.J. (1978*b*). Tolerance of tree roots to waterlogging: II. Adaptation of Sitka spruce and lodgepole pine to waterlogged soil. *New Phytologist*, **80**, 71–7.

Craib, I.J. (1939). *Thinning, pruning and management studies on the main exotic conifers grown in South Africa*. Department of Agriculture and Forestry Science Bulletin, 196. Government Printer, Pretoria.

Crawford, R.M.M. (1982). Physiological responses to flooding. In *Physiological plant ecology II* (ed. O.L. Lange, P.S. Nobel, C.B. Osmond, and H. Ziegler), pp. 454–77. Springer-Verlag, Berlin.

Cremer, J.W., Myers, B.J., Duys, F. van der, and Craig, I.E. (1977). Silvicultural lessons from the 1974 windthrow in radiata pine plantations near Canberra. *Australian Forestry*, **40**, 274–92.

Critchfield, W.B. (1957). Geographic variation in *Pinus contorta*. Maria Moors Cabat Foundation Publication 3. Harvard University Press.

Crooke, M. (1979). The development of populations of insects. In *Ecology of even-aged forest plantations* (ed. E.D. Ford, D.C. Malcolm, and J. Atterson), pp. 209–17. Institute of Terrestrial Ecology, Cambridge.

Crowe, S. (1978). *The landscape of forests and woods*. Forestry Commission Booklet, 44. HMSO, London.

Crowther, R.E. and Evans, J. (1984). *Coppice*, (2nd edn). Forestry Commission Leaflet, 83. HMSO, London.

Dallas, W.G. (1962). The progress of peatland afforestation in Northern Ireland. *Irish Forestry*, **19**, 84–93.

Dargavel, J. (ed.). (1990). *Prospects for Australian plantation forests*. Centre for Resource and Environmental Studies, Australian National University, Canberra.

Davidson, J. (1987). *Bioenergy tree plantations in the tropics: ecological impacts and their implications*. International Union for the Conservation of Nature Commission on Ecology, Paper No. 12. Gland, Switzerland.

Davies, R.J. (1984). The importance of weed control and the use of tree shelters for establishing broadleaved trees on grass-dominated sites in England. In proceedings *ECE/FAO/ILO Seminar Techniques and machines for the rehabilitation of low-productivity forest*. Turkey, May 1984.

Davis-Case, D. (1989). *Community forestry: participatory assessment, monitoring and evaluation*. Community Forestry Note 2. FAO, Rome.

Dawkins, H.C. (1958). The management of natural tropical high-forest with special reference to Uganda. Imperial Forestry Institute Oxford, Paper 34.

Dawkins, H.C. (1963). Crown diameters: their relation to bole diameter in tropical forest trees. *Commonwealth Forestry Review*, **42**, 318–33.

Dawson, W.M. and McCracken, A.R. (1995). The performance of polyclonal stands in short rotation coppice willow for energy production. *Biomass and Bioenergy*, **8**, 1–5.

Day, W.R. and Peace, T.R. (1946). *Spring frosts*. Forestry Commission Bulletin 18. HMSO, London.

Décourt, N. (1965). Le pin sylvestre et le pin laricio de Corse en Sologne. *Annales des Sciences Forestières*, **12**, 259–318.

Décourt, N. (1971). Tables de production provisoires pour l'épicéa commun dans le nord-est de la France. *Annales des Sciences Forestières*, **29**, 49–65.

Décourt, N. (1973). Tables de production pour l'épicéa commun et le Douglas dans l'ouest du Massif Central. *Revue Forestière Française*, **25**, 99–104.

Delabraze, P. (ed.) (1990*a*). Espaces forestiers et incendies. *Revue Forestière Française*, **42**, No. sp.

Delabraze, P. (1990*b*). Quelques concepts sylvicoles et principes d'aménagement, de prévention et de prévision du risque incendie. *Revue Forestière Française*, **42**, No. sp. 182–7.

Dickson, D.A. (1971). The effect of form, rate and position of phosphatic fertilizers on growth and nutrient uptake of Sitka spruce on deep peat. *Forestry*, **44**, 17–26.

Dickson, D.A. (1977). Nutrition of Sitka spruce on peat—problems and speculations. *Irish Forestry*, **34**, 31–9.

Dohrenbusch, A. and Frochot, H. (1993). Forest vegetation management in Europe. In *International Conference on Forest Vegetation Management Auburn Proceedings*, pp. 20–9.

Douglas, J.J. (1983). *A re-appraisal of forestry development in developing countries*. Martinus Nijhoff/ Dr W Junk, The Hague, Netherlands.

Douglas, J.J. (1986). Forestry and rural people: new economic perspectives. In *Proceedings, Division 4, 18th International Union of Forestry Research Organisations World Congress*, pp. 62–71. Ljubljana, Yugoslavia.

Drew, J.T. and Flewelling, J.W. (1977). Some recent Japanese theories of yield-density relationships and their application to Monterey pine plantations. *Forest Science*, **23**, 517–34.

Driessche, R. van den (1980). Effects of nitrogen and phosphorus fertilization on Douglas fir nursery growth and survival after outplanting. *Canadian Journal of Forest Research*, **10**, 65–70.

Driessche, R. van den (1982). Relationship between spacing and nitrogen fertilization of seedlings in the nursery, seedling size and outplanting performance. *Canadian Journal of Forest Research*, **12**, 865–75.

Driessche, R. van den (1983). Growth, survival and physiology of Douglas fir seedlings following root wrenching and fertilization. *Canadian Journal of Forest Research*, **13**, 270–8.

Driessche, R. van den (1984). Relationships between spacing and nitrogen fertilization of seedlings in the nursery, seedling mineral nutrition and outplanting performance. *Canadian Journal of Forest Research*, **14**, 431–6.

Du Boullay, Y. (1986). *Guide pratique du désherbage et du débroussaillement chimiques*. Institut pour le Développement Forestier, Paris.

Dubroca, E. (1983). Évolution saisonnière des réserves dans un taillis de châtaigniers, *Castanea sativa* Mill., avant et après la coupe. Thesis, Université de Paris–Sud.

Dupraz, C. (1994). Le chêne et le blé: l'agroforesterie peut-elle interesser les exploitations européennes des grandes cultures? *Revue Forestière Française*, **46**, No. sp. Agroforesterie en zone temperee, 84–95.

Dupraz, C., Guitton, J.L., Rapey, H., Bergez, J.E., and De Montard, F.X. (1993). Broad-leaved tree plantation on pastures : the treeshelter issue. In *Proceedings of the 4th International Symposium on Windbreaks and Agroforestry*, 106–111. Hedeselskabet, Viborg, Denmark.

Duvigneaud, P. (1985). *Le cycle biologique dans l'écosystème forêt*. ENGREF, Nancy, France.

Dykstra, D.P. (1984). *Mathematical programming for natural resource management*. McGraw-Hill, New York.

Edwards, P.N. (1980). Does pre-commercial thinning have a place in plantation forestry in Britain? In *Biologische, technische und wirtschaftliche Aspekte der Jungbestandflege* (ed. H. Kramer), Vol. 67, pp. 214–23. Schriften Forstlichen Fakultät, Universität Göttingen.

Edwards, P.N. and Christie, J.M. (1981). *Yield models for forest management*. Forestry Commission Booklet 48. (Plus numerous optional loose-leaf tables). Forestry Commission, Edinburgh, UK.

Effenterre, C. van (1990). Prévention des incendies de forêt; statistique et politiques. *Revue Forestière Française*, **42**, No. sp, 20–32.

Eichhorn, F. (1904). Beziehungen zwischen Bestandshöhe und Bestandsmasse. *Allgemeine Forst- und Jagdzeitung*, 45–9.

Eide, E. and Langsaeter, A. (1941). Produktions undersökelser i granskog. (Produktionsuntersuchungen von Fichtenwald). *Meddelelser fra Norsk Institutt for Skogforskning*, **26**. Oslo.

Eidmann, H.H. (1979). *Integrated management of pine weevil (Hylobius abietis L.) populations in Sweden*. USDAFS General Technical Report WO–8, pp. 103–9.

Eis, S. (1978). Natural root forms of western conifers. In *Proceedings of the root form of planted trees symposium* (ed. E. van Eerden and J.M. Kinghorn), pp. 23–7. British Columbia Ministry of Forests/Canadian Forestry Service Joint Report 8.

Ekö, P.M. (1985). [A growth simulator for Swedish forests based on data from the national forest survey]. Department of Silviculture, Swedish University of Agricultural Sciences, Umeå, Report 16.

Elfving, B. and Norgren, O. (1993). Volume yield superiority of lodpole pine compared to Scots pine in Sweden. In *Pinus contorta—from untamed forest to domesticated crop*. Department of Genetics and Plant Physiology, Swedish University of Agricultural Sciences, Umeå, Report 11, pp. 69–80.

Elliott, D.A., James, R.N., McLean, D.W., and Sutton, W.R.J. (1989). Financial returns from plantation forestry in New Zealand. In *Proceedings 13th Commonwealth Forestry Conference*, 6C. Rotorua, New Zealand.

Eriksson, H. (1976). [*Yield in Norway spruce in Sweden*]. Skogshög-skolan,Institutionen for skogsproduktion, Rapporter och Uppsatser No. 41. Stockholm.

Eriksson, H. and Johansson, U. (1993). Yields of Norway spruce in two consecutive rotations in southwestern Sweden. *Plant and Soil*, **154**, 239–47.

Étienne, M. (ed.) (1995). *Silvopastoral systems in the French Mediterranean region*. INRA, Avignon.

Etverk, I. (1972). [Factors affecting the resistance of stands to storms.] *Metsanduslikud Uurimused, Estorian SSR9*, 222–36.

Evans, H. (1996). Entomological threats to British forestry. *Institute of Chartered Foresters News*, **3**, 5–7.

Evans, J. (1976). Plantations: productivity and prospects. *Australian Forestry*, **39**, 150–63.

Evans, J. (1982*a*). Silviculture of oak and beech in northern France: observations and current trends. *Quarterly Journal of Forestry*, **76**, 75–82.

Evans, J. (1982*b*). *Sweet chestnut coppice*. Forestry Commission Research Information Note 770/82.

Evans, J. (1983). Choice of Eucalyptus species for cold temperate atlantic climates. In *Frost resistant eucalypts*. International Union of Forestry Research Organisations/Association Forêt-Cellulose Symposium, Bordeaux, France.

Evans, J. (1984). *Silviculture of broadleaved woodland*. Forestry Commission Bulletin, 62. HMSO, London.

Evans, J. (1992). *Plantation forestry in the tropics*. Clarendon Press, Oxford.

Evans, J. (1994). Long-term experimentation in forestry and site change. In *Long-term experiments in agricultural and ecological sciences* (ed. R.A. Leigh and A.E. Johnston), pp. 83–94. CAB International.

Evans, J. (1996). The sustainability of wood production from plantations: evidence over three successive rotations in the Usutu Forest, Swaziland. *Commonwealth Forestry Review*, **75**, 234–9.

Evans, J. and Hibberd, B.G. (1990). Managing to diversify forests. *Arboricultural Journal*, **14**, 373–8.

Evert, F. (1971). *Spacing studies—a review*. Canadian Forest Service. Forest Management Institute Information Report FMR–X–37.

Faber, P.J. and Burg, J. van den. (1982). [The production of woody biomass.] *Nederlands Bosbouw Tijdschrift*, **54**, 198–205.

Faber, P.J. and Sissingh, G. (1975). Stability of stands to wind. I. A theoretical approach. II. The practical viewpoint. *Nederlands Bosbouw Tijdschrift*, **47**, 179–93.

FAO (Food and Agriculture Organization) (1967). Actual and potential role of man-made forests in the changing world pattern of wood consumption. In *World symposium of man-made forests and their industrial importance*. Vol. 1, pp. 1–51. Canberra, Australia. FAO, Rome.

FAO (Food and Agriculture Organization) (1980). *Poplars and willows in wood production and land use*. FAO Forestry Series No. 10. FAO, Rome.

FAO (Food and Agriculture Organization) (1985). *Tree growing by rural people*. Forestry Paper 64. FAO, Rome.

Feeny, P. (1976). Plant apparency and chemical defense. In *Recent advances in phytochemistry* (ed. J.W. Wallace and R.L. Mansell), Vol. 10, pp. 1–40. Plenum Press, New York.

Felker, P. (1981). Uses of tree legumes in semi-arid areas. *Economic Botany*, **35**, 174–86.

Fenton, R.T. (1967). Rotations in man-made forests. *In World symposium on man-made forests and their industrial importance*. Vol. 1. Canberra, Australia. FAO, Rome.

Ferris-Kaan, R. (ed.) (1995). *The ecology of woodland creation*. Wiley & Sons, Chichester.

Finnigan, J.J. and Brunet, Y. (1995). Turbulent airflow in forest on flat and hilly terrain. In *Wind and trees* (ed. M. Coutts and J. Grace). Cambridge University Press.

Ford, E.D. (1980). Can we design a short rotation silviculture for windthrow-prone areas? In *Research strategy for silviculture* (ed. D.C. Malcolm), pp. 25–34. Institute of Foresters of Great Britain, Edinburgh.

Ford, E.D. (1982). Catastrophe and disruption in forest ecosystems and their implications for plantation forestry. *Scottish Forestry*, **36**, 9–24.

Ford-Robertson, F.C. (1971). *Terminology of forest science, technology, practice and products*. Multilingual Forestry Terminology Series No. 1. Society of American Foresters, Washington, DC.

Forestry Commission (1993). *Forests and water. Guidelines* (3rd edn). HMSO, London.

Forestry Commission (1994). *Forest landscape design. Guidelines* (2nd edn). HMSO, London.

Forestry Commission (1995). *Forests and archaeology: Guidelines*. HMSO, London.

Forestry Commission of Tasmania (1989). *Forest practices code*. Forestry Commission of Tasmania, Hobart, Tasmania.

Forestry Department, Brisbane (1963). *Technique for the establishment and maintenance of plantations of hoop pine*, pp. 22–8. Government Printer, Brisbane.

Forestry Department, Queensland (1981). Exotic conifer plantations. In *Research Report 1981*, pp. 37–41. Department of Forestry, Brisbane, Queensland.

Formby, J. (1986). *Approaches to social impact assessment*. Working Paper 1986/8, Centre for Resource and Environmental Studies, Australian National University, Canberra.

Fox, J.E.D. (1984). Rehabilitation of mined lands. *Forestry Abstracts* (review article), **45**, 565–600.

Fox, J.F. (1983). Post-fire succession of small-mammal and bird communities. In *The role of fire in northern circumpolar ecosystems* (ed. R.W. Wein and D.A. MacLean), pp. 155–80. Wiley, New York.

Franklin, E.C. (1989). Selection strategies for eucalypt tree improvement—four generations of selection in *Eucalyptus grandis* demonstrates valuable methodology. In *Breeding tropical trees* (ed. G.L. Gibson, A.R. Griffin, and A.C. Matheson), pp. 197–209. Oxford Forestry Institute and Winrock International.

Fraser, A.I. (1964). Wind tunnel and other related studies on coniferous trees and tree crops. *Scottish Forestry*, **18**, 84–92.

French, D.W. and Schroeder, D.B. (1969). The oak wilt fungus, *Ceratocystis fagacearum*, as a selective silvicide. *Forest Science*, **15**, 198–203.

Frochot, H. (1988). Techniques particulières de reboisement sur les délaissés: concurrence avec la végétation herbacée. In *19ème congrès de l'union européenne des forestiers*, Nîmes-Nancy, pp. 65–6.

Frochot, H. (1990). Weed control in French forests. In *Tendencias mundiales en el control de la vegetacion accesoria en los montes*. Universidad politecnica de Madrid and Asociacion ingenieros montes, Madrid.

Frochot, H. (1992). Economic and ecological aspects of forest vegetation management in France. In *International Union of Forestry Research Organisations Centennial*, p. 260. Berlin.

Frochot, H. and Lévy, G. (1986). Facteurs du milieu et optimisation de la croissance initiale en plantations de feuillus. *Revue Forestière Française*, **38**, 301–6.

Frochot, H. and Trichet, P. (1988). Influence de la compétition herbacée sur la croissance de jeunes pins sylvestres. In *8ème colloque international sur la biologie, l'écologie et la systématique des mauvaises herbes*, annales A.N.P.P. No. 3, pp. 509–15.

Frochot, H., Dohrenbusch, A., and Reinecke, H. (1990). Forest weed management: recent developments in France and Germany. In *International Union of Forestry Research Organisations 19th world congress*, Montreal, pp. 290–9.

Frochot, H., Lévy, G., Lefèvre, Y., and Wehrlen, L. (1992). Amélioration du démarrage des plantations de feuillus précieux: cas du frêne en station à bonne réserve en eau. *Revue Forestière Française*, **44**, 61–5.

Fryer, J.D. and Makepeace, R.J. (1977). *Weed control handbook*, Vol. 1: *Principles* (6th edn). Blackwell, Oxford.

Fu, X.K. (1993). The effect of shelterbelt net on biomass of crops and vegetation in Northern China. In *Windbreaks and agroforestry*, pp. 47–9. Hedeselskabet, Viborg, Denmark..

Funk, D.T. (1979). Stem form response to repeated pruning of young black walnut trees. *Canadian Journal of Forest Research*, **9**, 114–16.

Gama, A., Frochot, H., and Delabraze, P. (1987). *Phytocides en sylviculture*, Note technique No. 53. CEMAGREF, Nogent-sur-Vernisson.

Garbaye, J. (1990). Pourquoi et comment observer l'état mycorhizien des plants forestiers? *Revue Forestière Française*, **42**, 35–47.

Garbaye, J. (1991). Les mycorhizes des arbres et plantes cultivées. In *Lavoisier* (ed. D.G. Strullu) , pp. 197–248. Paris.

Gardiner, B. and Stacey, G. (1996). *Designing forest edges to improve wind stability*. Forestry Commission Technical Paper 16.

Gauthier, J.J. (1991). Les bois de plantation dans le commerce mondial des produits forestiers. In *L'emergence des nouveaux potentiels forestiers dans le monde*, pp. 9–20.Association Forêt-Cellulose, Paris.

Gerischer, G.F.R. and Villiers, A.M. de (1963). The effect of heavy pruning on timber properties. *Forestry in South Africa*, **3**, 15–41.

Gibson, I.A.S. and Jones, T. (1977). Monoculture as the origin of major forest pests and diseases. In *Origins of pest, parasite, disease and weed problems* (ed. J.M. Cherritt and G.R. Sagar), pp. 139–61. Blackwell, Oxford.

Gibson, I.A.S., Burley, J., and Speight, M.R. (1982). The adoption of agricultural practices for the development of heritable resistance to pests and pathogens in forest crops. In *Resistance to diseases and pests in forest trees* (ed. H.M. Heybroek, B.R. Stephan, and J. von Weissenberg), pp. 9–21. Centre for Agricultural Publishing and Documentation, Wageningen, Netherlands.

Giertych, M.M. (1976). Summary results of the International Union of Forestry Research Organisations 1938 Norway spruce provenance experiment height growth. *Silvae Genetica*, **25**, 154–64.

Gill, C.J. (1970). The flooding tolerance of woody species—a review. *Forestry Abstracts* (review article), **31**, 671–88.

Gilmour, D.A., King, G.C., and Hobley, M. (1989). Management of forests for local use in the hills of Nepal. 1. Changing forest management paradigms. *Journal World Forest Resource Management*, **4**, 93–110.

Gilmour, D.A., King, G.C., Applegate, G.B., and Mohns, B. (1990). Silviculture of plantation forests in central Nepal to maximize community benefits. *Forest Ecology and Management*, **32**, 173–86.

Gondard, P. (1988). Land use in the Andean region of Ecuador. *Land Use Policy*, **65**, 341–8.

Goor, C.P. van (1970). Fertilization of conifer plantations. *Irish Forestry*, **27**, 68–80.

Gordon, G.T. (1973*). Damage from wind and other causes in mixed white fir-red fir stands adjacent to clear cuttings*. Research Paper Pacific Southwest Forest and Range Experiment Station PSW–90. USDA Forest Service.

Goulet, F. (1995). Frost heaving of forest tree seedlings: a review. *New Forests*, **9**, 67–94.

Grace, J. (1977). *Plant responses to wind*. Academic Press, London.

Grace, J. (1983). *Plant–atmosphere relationships*. Chapman and Hall, London.

Grassi, G., Delmon B., Molle J.F., and Zibetta H. (ed.) (1987). *Biomass for energy and industry*, 4th EC conference. Elsevier, London.

Grayson, A.J. (1993). *Private forest policy in western Europe*. CAB International, Wallingford, UK.

Gregori, G.L. and Tocci, A. (1990). La trufficulture dans les pays méditerranéens: l'exemple de l'Italie. *Forêt Méditerranéenne*, **12**, 143–52.

Greig, B.J.W. (1995). Butt-rot of Scots pine in Thetford Forest caused by *Heterobasidion annosum*: a local phenomenon. *European Journal of Forest Pathology*, **25**, 95–9.

Grieve, I.C. (1978). Some effects of the plantation of conifers on a freely drained lowland soil, Forest of Dean, UK. *Forestry*, **51**, 21–8.

Gryse, J.J. de (1955). *Forest pathology in New Zealand*. New Zealand Forest Service Bulletin, 11.

Guitton, J.L. (ed.) (1994). Agroforesterie en zone tempérée. *Revue Forestière Française*, 46.

Guyot, G., Ben Salem, B., and Delecolle, R. (1986). *Brise–vent et rideaux–abris avec référence particulière aux zones sèches*. Cahier FAO conservation No.15. FAO, Rome.

Habjørg, A. (1971). *Effects of photoperiod on temperature growth and development of three longitudinal and three altitudinal populations of Betula pubescens*. Meldinger fra Norges landbrukshogskole, 51.

Hagen, L.J. (1976). Windbreak design for optimum wind erosion control. In *Shelterbelts on the Great Plains*. In Great Plains Agricultural Publication, (ed. R.W. Tinus), 78, pp. 31–6.

Hägglund, B. (1974). *Site index curves for Scots pine in Sweden*. Rapporter och Uppsatser, No. 31.

Hägglund, B. (1981). *Forecasting growth and yield in established forests*. Department of Forest Measuration, Swedish University of Agricultural Siences, Umeå, Report 31.

Hall, D.O. (1983). Food versus fuel, a world problem? In *Energy from Biomass* (ed. A. Strub, P. Chartier, and G. Schleser), pp. 43–62. Applied Science Publishers, London.

Hämälänen, J. (1990). *Advantages of different site preparation methods on planting results and costs in southern Finland*. Metsäteho report 402.

Hamilton, G.J. (1976a). The Bowmont Norway spruce thinning experiment 1930–1974. *Forestry*, **49**, 109–19.

Hamilton, G.J. (ed.) (1976b). Effects of line thinning on increment. In *Aspects of thinning*. Forestry Commission Bulletin, **55**, pp. 37–45. HMSO, London.

Hamilton, G.J. (1980). *Line thinning*. Forestry Commission Leaflet, 77. HMSO, London.

Hamilton, G.J. (1981). The effect of high intensity thinning on yield. *Forestry*, **54**, 1–15.

Hamilton, G.J. and Christie, J.M. (1971). *Forest management tables (metric)*. Forestry Commission Booklet, 34. HMSO, London.

Hånell, B. (1988). Postdrainage forest productivity of peatlands in Sweden. *Canadian Journal of Forest Research*, **18**, 1443–56.

Hansen, E.A. (1988). SRIC yields: a look to the future. In *Economic evaluations of short rotation biomass energy systems*. IEA Report 88 (2). pp. 197–207. Forest products laboratory, USDA Forest Service, Madison, Wis, USA.

Harley, J.L. (1971). Fungi in ecosystems. *Journal of Applied Ecology*, **8**, 627–42.

Harley, L.J. and Smith, S.E. (1983). *Mycorrhizal symbioses*. Academic Press, London.

Harper, J.L. (1977). *Population biology of plants*. Academic Press, London.

Harper, J.L. (1982). The concept of population in modular organisms. In *Theoretical ecology* (2nd edn) (ed. R.M. May), pp. 53–77. Blackwell, Oxford.

Heal, O.W., Swift, M.J., and Anderson, J.M. (1982). Nitrogen cycling in United Kingdom forests: the relevance of basic ecological research. *Philosophical Transactions of the Royal Society Series*, **B296**, 427–44.

Heikurainen, L. (1979). Peatland classification in Finland and its utilization for forestry. In *Classification of peat and peatlands*. International Peat Society, Helsinki, pp. 135–46.

Heikurainen, L. and Parkarinen, P. (1982). Mire vegetation and site types. In *Peatlands and their utilzation in Finland* (ed. J. Laine). Finnish Peatland Society, International Peat Society, Finnish National Committee of the International Peat Society, Helsinki, pp. 14–23.

Heilman, P., Dao, T., Cheng, H.H., Webster, S.R., and Christensen, L. (1982*a*). Comparison of fall and spring applications of 15N-labelled urea to Douglas fir: II. Fertilizer nitrogen recovery in trees and soil after two years. *Soil Science Society of America Journal*, **46**, 1300–4.

Heilman, P., Dao, T., Cheng, H.H., Webster, S.R., and Harper, S.S. (1982*b*). Comparison of fall and spring applications of 15N-labelled urea to Douglas fir: I. Growth responses and nitrogen levels in foliage and soil. *Soil Science Society of America Journal*, **46**, 1293–9.

Helles, F. and Linddal, M. (1996). *Afforestation in Nordic countries*. The Nordic Council 1996:15, Copenhagen.

Hellum, A.J. (1978). The growth of planted spruce in Alberta. In *Proceedings of the root form of planted trees symposium* (ed. E. van Eerden and J.M. Kinghorn), pp. 191–6.

Hengst, E. and Schulze, W. (1976). [Examples of spatial organization as a means of increasing the security of production in *Picea abies* forests.] *Wissenschaftliche Zeitschrift der Technischen Universität*, Dresden, **25**, 313–14.

Henman, D.W. (1963). *Pruning conifers for the production of quality timber*. Forestry Commission Bulletin, 35. HMSO, London.

Heybroek, H.M. (1981). *Possibilities of clonal plantations in the 1980s and beyond*. Commonwealth Forestry Institute Occasional Papers, 15 (ed. J A. Longman), pp. 21–2.

Hibberd, B.G. (1985). Restructuring of plantations in Kielder Forest District. *Forestry*, **58**, 119–29.

Hibberd, B.G. (1989). *Farm woodland practice*. Forestry Commission Handbook, 3. HMSO, London.

Hibberd, B.G. (1991). *Forestry Practice*. Forestry Commission Handbook, 6. HMSO, London.

Higuchi, T., Ito, T., Umezawa, T., Hibino, T., and Shibata, D. (1994). Red-brown color of lignified tissues of transgenic plants with antisense CAD gene: wine-red lignin from coniferyl aldehyde. *Journal of Biotechnology*, **37**, 151–8.

Hill, H.W. (1979). Severe damage to forests in Canterbury, New Zealand resulting from orographically reinforced winds. In *Symposium of forest meteorology*, pp. 22–40. World Meteorological Organization 527. Canadian Forest Service, Canada.

Hinson, W.H., Pyatt, D.G., and Fourt, D.F. (1970). Drainage studies. In *Report on forest research 1970*, pp. 90–1. Forestry Commission, HMSO, London.

Hodge, S.J. (1995). *Creating and managing woodland around towns*. Forestry Commission Handbook, 11. HMSO, London.

Holloway, C.W. (1967). The protection of man-made forests from wildlife. In *FAO World symposium on man-made forests and their industrial importance*, Vol. 1, pp. 697–715. Canberra, Australia.

Holmes, G.D. (1980). Weed control in forestry—achievements and prospects in Britain. In *Proceedings of the conference on weed control in forestry*, 1–2 April 1980, University of Nottingham, pp. 1–11.

Holmsgaard, E., Holstener-Jørgensen, H., and Yde-Andersen, A. (1961). [*Soil formation, increment and health of first- and second-generation stands of Norway spruce*.]. Forstlige Forsøgsvaesen i Danmark, 27 (1).

Holstener-Jørgensen, H. (1977). Plant nutrient balance in decoration greenery cultivation. *Silvae Fennica*, **1**, 230–3.

Hubert, M. and Courraud, R. (1987). *Elagage et taille de formation des arbres forestiers*. Institute pour le Développement Forestier, Paris.

Hugon, D. (1994). L'élagage à grande hauteur. *Silva Belgica*, **101**, 35–40.

Hummel, F.C. (1991). Comparisons of forestry in Britain and mainland Europe. *Forestry*, **64**, 141–55.

Hummel, F.C., Palz, W., and Grassi, G. (ed.) (1988). *Biomass forestry in Europe*: *A strategy for the future*. Elsevier, London.

Hutchinson, C.E. (1965). *The ecology theater and the evolutionary play*. Yale University Press, Newhaven, CT, USA.

Hütte, P. (1968). Experiments on windthrow and wind damage in Germany: site and susceptibility of spruce forests to storm damage. *Forestry*, **41**, (suppl.), 20–6.

Hüttl, R.F., Nilsson, L.O., and Johansson, U.T. (ed.) (1995). Nutrient uptake and cycling in forest ecosystems. *Plant and Soil*, **168–169**. Many papers.

Huuri, O. (1978). Effect of various treatments at planting and of soft containers in the development of Scots pine. In *Proceedings of the root form of planted trees symposium* (ed. E. van Eerden and J.M. Kinghorn), pp. 101–18. British Columbia Ministry of Forests/Canadian Forestry Service Joint Report 8.

Hyman, E.L. and Stiftel, B. (1988). *Combining facts and values in environmental impact assessment.* Social Impact Assessment Series No. 16. Westview Press, Boulder, CO, USA.

IDF (1981). *La réalisation pratique des haies brise–vent et bandes boisées.* Institut pour le Développement Forestier, Paris.

Incoll, L.D., Corry, D.T., Wright, C., Hardy, D., Compton, S.G., Naeem, M., and Klaa, K. (1995). Silvoarable experiment with quality timber production hedges. *Agroforestry Forum*, **5**, 14–15.

Ingestad, T. (1991). Nutrition and growth of forest trees. *Tappi Journal*, **74**, 55–62.

Innes, J.L. (1993). *Forest health: its assessment and status.* CAB International, Wallingford, UK.

Insley, H. (1982). The influence of post planting maintenance on the growth of newly planted broadleaved trees. In *Proceedings of Conference of Horticultural Education Association*, 5–8th April 1982 Bridgewater, UK, pp. 74–80.

Insley, H. (1988). *Farm woodland planning.* Forestry Commission Bulletin, 80. HMSO, London.

ITTO (1990). *ITTO guidelines for the sustainable management of natural tropical forests.* International Tropical Timber Council, Technical Series, 5. Yokohama, Japan.

ITTO (1991). *Draft report of Working Group on ITTO guidelines for the establishment and sustainable management of planted tropical forests.* International Tropical Timber Council, Yokohama.

Jack, W.H. (1965). Experiments on tree growing on peat in N. Ireland. *Forestry*, **38**, 220–40.

Jane, G.T. (1986). Wind damage as an ecological process in mountain beech forests of Canterbury, New Zealand. *New Zealand Journal of Ecology*, **9**, 25–39.

Jarvis, P.G. and Leverenz, J.W. (1983). Productivity of temperate, deciduous and evergreen forests. In *Physiological Plant Ecology* IV (ed. O.L. Lange, P.S. Nobel, C.B. Osmond, and M. Ziegler), pp. 233–80. Springer-Verlag, Berlin.

Jinks, R.L. (1994). Container production of tree seedings. In *Nursery practice* (ed. J.R. Aldhous and W.L. Mason), pp. 122–34. Forestry Commission Bulletin, 111. HMSO, London.

Jobling, J. (1990). *Poplars for wood production and amenity.* Forestry Commission Bulletin, **92**. HMSO, London.

Johnson, V.J. (1984). Prescribed burning. *Journal of Forestry*, **82**, 82–90.

Johnston, D.R., Grayson, A.J., and Bradley, R.T. (1967). *Forest planning.* Faber and Faber, London.

Jones, E.W. (1965). Pure conifers in central Europe—a review of some old and new work. *Journal of the Oxford University Forestry Society*, **13**, 3–15.

Jones, H.E., Quarmby, C., and Harrison, A.F. (1991). A root bioassay test for nitrogen deficiency in forest trees. *Forest Ecology and Management*, **42**, 267–82.

Jørgensen, K. (1993). Legislation in Denmark in 100 years, or aid scheme for shelterbelt planting in Denmark. In *Windbreaks and agroforestry*, pp. 123–5. Hedeselskabet, Viborg, Denmark.

Jouanin, L., Brasileiro, A.C.M., Lepié, J.C., Pilate, G., and Cornu, D. (1993). Genetic transformation: a short review of methods and their applications, results and perspectives for forest trees. *Annales des Sciences Forestières*, **50**, 325–36.

Kanowski, P.J. (1995). The complex future of plantation forestry. In *CRC for Temperate Hardwood Forestry*, pp. 483–7. International Union of Forestry Research Organisations, Hobart, Tasmania.

Kanowski, P.J. and Savill, P.S. (1992). Forest plantations: towards sustainable practice. In *Plantation politics* (ed. C. Sargent and S. Bass), pp. 121–55. Earthscan Publications, London.

Kanowski, P.J., Savill, P.S., Adlard, P.G., Burley, J., Evans, J., Palmer, J.R., and Wood, P.J. (1992). Plantation forestry. In *Managing the world's forests* (ed. N.P. Sharma), pp. 375–402. Kendall/Hunt, Iowa, USA.

Karim, A.B. and Savill, P.S. (1991). Effect of spacing on growth and biomass production of *Gliricidia sepium* in an alley cropping system in Sierra Leone. *Agroforestry Systems*, **16**, 213–22.

Karlman, M. (1981). The introduction of exotic tree species with special reference to *Pinus contorta* in northern Sweden, review and background. *Studia Forestalia Suecica*, **158**, 1–25.

Karlman, M., Hansson, P., and Witzell, J. (1994). *Scleroderris* canker on lodgepole pine introduced in northern Sweden. *Canadian Journal of Forest Research*, **24**, 1948–59.

Keller, R. and Thiercelin, F. (1984). L'élagage des plantations d'épicéa commun et de Douglas. *Revue Forestière Française*, **36**, 289–301.

Kendrew, W.G. (1961). *The climates of continents*. Clarendon Press, Oxford.

Kennedy, C.J.E. and Southwood, T.R.E. (1984). The number of species associated with British trees. *Journal of Animal Ecology*, **53**, 455–78.

Keresztesi, B. (1983). Breeding and cultivation of black locust in Hungary. *Forest Ecology and Management*, **6**, 217–44.

Kerr, G. and Evans, J. (1993). *Growing broadleaves for timber*. Forestry Commission Handbook, 9. HMSO, London.

Kerr, G. and Jinks, R.L. (1994). A comparison of cell-grown and bare-rooted oak and beech seedlings one season after out-planting. *Forestry*, **67**, 297–312.

Khanna, P.J. (1981). Soil analyses for evaluation of forest nutrient supply. In *Proceedings of Australian forest nutrition workshop: productivity in perpetuity*, Australia, 10–14 August 1981, pp. 231–8. CSIRO Division of Forest Research.

Kilgore, B.M. (1987). The role of fire in wilderness: a state-of-knowledge review. In *Proceedings National Wilderness Research Conference: issues, state-of-knowledge, future directions* (ed R.C. Lucas), pp. 70–103. US Forest Service General Technical Report INT–220..

Kilian, W. (1981). Site classification systems used in forestry. In *Proceedings of the workshop on land evaluation for forestry* (ed. P. Laban), pp. 134–51. International Institute for Land Reclamation and Improvement Publication 28. Wageningen, The Netherlands.

Kilpatrick, D.J., Sanderson, J.M., and Savill, P.S. (1981). The influence of five early respacing treatments on the growth of Sitka spruce. *Forestry*, **54**, 17–29.

Kira, T. (1975). Primary production in forests. In *Photosynthesis and productivity in different environments* (ed. J.P. Cooper), pp. 5–40. Cambridge University Press.

Kira, T. and Shidei, T. (1967). Primary production and turnover of organic matter in different forest ecosystems of western Pacific. *Japanese Journal of Ecology*, **17**, 80–7.

Kirkland, A. (1989). The rise and fall of multiple-use management in New Zealand. *New Zealand Forestry*, **33**, 9–12.

Kivinen, E. and Pakarinen, P. (1981). Geographical distribution of peat resources and major peatland complex types in the world. *Ann Acad Sci Fenn Ser A III Geol Geogr*, **132**, 1–28.

Kleinschmit, J. (1986). État actuel et perspectives d'avenir de l'amélioration des arbres forestiers. *Revue Forestière Française*, **38**, No. sp. *Amélioration génétique des arbres forestiers*, 198–200.

Kleinschmit, J. and Otto, H.–J. (1974). [Rehabilitation of the gale-damage of forests of Lower Saxony.] *Forst und Holzwirt*, **29**, 1–12.

Knopp, T.B. and Caldbeck, E.S. (1990). The role of participatory democracy in forest management. *Journal of Forestry*, **88**, 13–18.

Knowles, R.L. (1991). New Zealand experience with silvopastoral systems: a review. *Forest Ecology and Management*, **45**, 251–67.

König, E. and Gossow, H. (1979). Even-aged stands as habitat for deer in central Europe. In *Ecology of even-aged forest plantations* (ed. E.D. Ford, D.C. Malcolm, and J. Atterson), pp. 429–51. Institute of Terrestrial Ecology, Cambridge.

Koster, R. (1981). [The cultivation of 'energy forest' in Sweden.] *Nederlands Bosbouw Tijdschrift*, **54**, 206–13.

Kozlowski, T.T. and Ahlgren, C.F. (ed.) (1974). *Fire and ecosystems*. Academic Press, New York.

Kramer, H. (1980). Tending and stability of Norway spruce stands. In *Stability of spruce ecosystems*, pp. 121–33. University of Agriculture, Brno, Czechoslovakia.

Kramer, H. and Bjerg, N. (1978). [*Biological aspects of tending young stands of Norway spruce.*] Forestry Faculty, University of Göttingen, West Germany, No. 55.

Kramer, H. and Spellmann, H. (1980). *Beitrage zur Bestandesbegrundung der Fichte*. Forestry Faculty University of Göttingen, No. 64.

Kremer, A., Savill, P.S., and Steiner, K.C. (1993). Genetics of oaks. *Annales des Sciences Forestières*, **50**, suppl. 1.

Kroth, W., Loffler, H.D., Plochman, R., and Rader-Roitch, J.E. (1976). *Forestry problems and their implications for the environment in member states of the EC.* Study PE 168 Vol. III, Munich, West Germany.

Labaznikov, B.V. (1982). [Geographical variation in the protein content of grain crops in fields protected by shelterbelts.] *Lesnoe Khozaistvo,* **8**, 30–1. (Cited from *Forestry Abstracts* (1983) **44**, FA6786).

Label, P., Sotta, B., and Miginiac, E. (1989). Endogenous levels of abscisic acid and indole-3-acetic acid during in vitro rooting of wild cherry explants produced by micropropagation. *Plant Growth Regulation*, **8**, 325–33.

Lagercrantz, U. and Ryman, N. (1990). Genetic structure of Norway spruce: concordance of morphological and allozymic variation. *Evolution*, **44**, 38–53.

Lähde, F. (1969). Biological activity in some natural and drained peat soils with special reference to oxidation-reduction conditions. *Acta Forestalia Fennica*, **94**.

Lancaster, D.F., Walters, R.S., Laing, F.M., and Foulds, R.T. (1974). *A silvicultural guide for developing a sugarbush.* USDA Forest Service Research Paper NE–286.

Lanier, L. (1986). *Précis de sylviculture.* ENGREF, Nancy, France.

Leaf, A.L., Rathakette, P., and Solan, F.M. (1978). Nursery seedling quality in relation to plantation performance. In *Proceedings of the root form of planted trees symposium* (ed. E. van Eerden and J.M. Kinghorn), pp. 45–51.

Le Tacon, F. and Bouchard, D. (1991). Les possibilités de mycorhization contrôlée en sylviculture rempérée. *Forêt Entreprise*, **74**, 29–41.

Le Tacon, F., Garbaye, J., Bouchard, D., Chevalier, G., Olivier, J.M., Guimberteau, J., Poitou, N., and Frochot, H. (1988). Field results from ectomychorrhizal inoculation in France. In *Proceedings of the Canadian workshop on mycorrhizae in forestry* (ed. M. LaConde and Y. Piché), pp. 57–74. Université Laval, Québec.

Leban, J.M., Houllier, F., Goy, B., and Colin, F. (1991). La qualité du bois d'Épicéa commun en liaison avec les conditions de croissance. *Forêt Entreprise*, **80**, 11–27.

Lee, S.J. (1990). *Potential gains from genetically improved Sitka spruce.* Research Information Note 190. Forestry Commission, Edinburgh.

Lees, J.C. (1972). Soil aeration response to drainage intensity in basin peat. *Forestry*, **45**, 135–43.

Lefèvre, F., Villar, M., and Bonduelle, P. (1994). Peupliers. In *L'amélioration génétique des essences forestières. Forêt Entreprise*, **96**, 76–8.

Lelu, M.A., Bastien, C., Klimaszewska, K., Ward, C., and Charest, P.J. (1994*a*). An improved method for somatic plantlet production in hybrid larch (*Larix × leptoeuropaea*). *Plant Cell, Tissue and Organ Culture*, **36**, 107–27.

Lelu, M.A., Klimaszewska, K., and Charest, P.J. (1994*b*). Somatic embryogenesis from immature and mature zygotic embryos and from cotyledons and needles of somatic plantlets of *Larix. Canadian Journal of Forest Research*, **24**, 100–6.

Lembcke, G., Knapp, E., and Dittmar, O. (1981) Die neue DDR-Kiefern-ertragstafel 1975. *Beitrage fur die Forstwirtschaft*, **15**, 55–64.

Leslie, A.J. (1987). Economic feasibility of natural management of tropical forests. In *Natural management of tropical moist forests* (ed. F.R. Mergen and J.R. Vincent), pp. 177–98. School of Forestry and Environmental Studies, Yale University, New Haven, CT, USA.

Leuschner, W.A. (1984). *Introduction to forest resource management*. Wiley, New York.

Levitt, J. (1972). *Responses of plants to environmental stress*. Academic Press, New York.

Lévy, G., Frochot, H., and Becker, M. (1990). Installation des peuplements de chênes et facteurs du milieu. *Revue Forestière Française*, **42**.

Leyton, L. (1958). The mineral requirements of forest plants. *Handbuch der Pflanzenphysiologie*, **6**, 1026–39.

Leyton, L. (1972). Forests, flooding and soil moisture. In *Proceedings Piene: Loro Previsione e difesa del suolo*, Rome, 23–30 November 1969, 327–37. Accademia Nazionale dei Lincei, 1972.

Leyton, L. (1975). *Fluid behaviour in biological systems*. Oxford University Press.

Li, C.Y., Lu, J.C., Trappe, J.M., and Bollen, W.B. (1967). Selective nitrogen assimilation by *Poria weiru*. *Nature*, **213**, 814.

Likens, G.E. and Bormann, F.H. (1995). *Biogeochemistry of a forested ecosystem*, (2nd edn). Springer-Verlag, New York.

Lindholm, T. and Vasander, H. (1987). Vegetation and stand development of mesic forest after prescribed burning. *Silva Fennica*, **21**, 259–78.

Lines, R. (1967). *The planning and conduct of provenance experiments*. Forestry Commission Research and Development Paper, 45. HMSO, London.

Lines, R. (1984). Species and seed origin trials in the industrial Pennines. *Quarterly Journal of Forestry*, **78**, 9–13.

Lines, R. (1985). The Macedonian pine in the Balkans and Great Britain. *Forestry*, **58**, 27–40.

Lines, R. (1987). *Choice of seed origins for the main forest species in Britain*. Forestry Commission Bulletin, 66. HMSO, London.

Lines, R. (1996). *Experiments on lodgepole pine seed origins in Britain*. Forestry Commission Technical Paper, 10.

Litvina, I.V. and Takle, E.S. (1993). Designing shelterbelts by use of a turbulence model. In *Windbreaks and agroforestry* (ed. Hedeselskabet), pp. 24–7. Viborg, Denmark.

Lorimer, C.G. (1977). The presettlement forest and natural disturbance cycle of northeastern Maine. *Ecology*, **58**, 130–48.

Lucas, O.W.R. (1983). *Design of landform and planting*. Forestry Commission Research and Development Paper, 132, pp. 24–36.

Luscher, P., Pelissier, D., and Bartoli, M. (1991). *L'élagage des résineux de l'Aude. Études technique et économique*. Bulletin technique, 19, pp. 35–44. Office National des Forêts, Paris.

MacLaren, P. (1983). Chemical welfare in the forest: a review of allelopathy with regard to New Zealand. *New Zealand Journal of Forestry*, **28**, 73–92.

Malcolm, D.C. (1979). The future development of even-aged plantations: silvicultural implications. In *Ecology of even-aged forest plantations* (ed. E.D. Ford, D.C. Malcolm, and J. Atterson), pp. 481–504. Institute of Terrestrial Ecology, Cambridge.

Marrs, R.H., Owen, L.D.C., Roberts, R.D., and Bradshaw, A.W. (1982). Tree lupin an ideal nurse crop for land restoration and amenity plantings. *Arboricultural Journal*, **6**, 161–74.

Marx, D.H. (1977). The role of mycorrhizae in forest production. *TAPPI conference*, Annual meeting. Atlanta, pp. 151–61.

Maser, C. (1990). *The redesigned forest*. Stoddart, Toronto.

Mason, W.L. (1992). Reducing the cost of Sitka spruce cuttings. In *Super Sitka for the 90s*. Forestry Commission Bulletin, 103, pp. 25–41. HMSO, London..

Mason, W.L. and Jinks, R.L. (1994). Vegetative propagation. In *Nursery practice* (ed. J.R. Aldhous and W.L. Mason), pp. 135–47. Forestry Commission Bulletin, 111. HMSO, London.

Matthews, J.D. (1989). *Silvicultural systems*. Oxford University Press.

Maugé, J.-P. (1986). La culture moderne du pin maritime dans les landes de Gascogne. *Forêt–Enterprise*, **37**, 1–28.

Maugé, J.-P. (1987). *Le pin maritime*. Institut pour le Développement Forestier, Paris.

Mayer, H. (1989). Windthrow. *Philosophical Transactions of the Royal Society of London*, **B324**, 267–81.

Mayhead, G.J. (1973). Some drag co-efficients for British forest trees derived from wind tunnel studies. *Agricultural Meteorology*, **12**, 123–30.

Mayhead, G.J., Gardiner, J.B.H., and Durrant, D.W. (1975). *A report on the physical properties of conifers in relation to plantation stability*. Forestry Commission Research and Development Division Paper (unpublished).

McAllister, J.S.V. and Savill, P.S. (1977). Effects of pig and cow slurry on the growth of Sitka spruce on oligotrophic peat and gley soils in Northern Ireland. *Irish Forestry*, **34**, 77–84.

McCracken, A.R. and Dawson, W.M. (1996). Interaction of willow (*Salix*) clones grown in polyclonal stands in short rotation coppice. *Biomass and Bioenergy*, **10**, 307–11.

McDonald, M.A., Malcolm, D.C., and Harrison, A.F. (1991). The use of ^{32}P root bioassay to indicate the phosphorus status of forest trees. *Canadian Journal of Forest Research*, **21**, 1180–93.

McElroy, G.H. (1981). Energy from biomass/novel sources of cellulose. In *Annual Report Loughgall Horticultural Centre,* pp. 68–71. Department of Agriculture, Northern Ireland,.

McGaughey, S.E. and Gregersen, H.M. (1988). *Investment policies and financing mechanisms for sustainable forestry development*. Inter-American Development Bank, Washington, DC.

McIntosh, R. (1983). In *Forestry Commission Report on Forest Research 1983*. HMSO, London.

McKay, H.M., Aldhous, J.R., and Mason, W.L. (1994). Lifting, storage, handling and despatch. In *Nursery practice* (ed. J.R. Aldhous and W.L. Mason), pp. 198–222. Forestry Commission Bulletin, 111. HMSO, London.

McMahon, T.A. (1975). The mechanical design of trees. *Scientific American*, **233**, 92–102.

Mengel, K. and Kirkby, E.A. (1978). *Principles of plant nutrition*. International Potash Institute, Bern, Switzerland.

Meredieu, C., Arrouays, D., Goulard, M., and Auclair, D. (1996). Short range soil variability and its effect on red oak growth (*Quercus rubra* L.). *Soil Science*, **161**, 29–38.

Miles, J. (1981). Effects of trees on soils. In *Forest and woodland ecology* (ed. F.T. Last and A.S. Gardiner), pp. 85–8. Institute of Terrestrial Ecology, Cambridge.

Miller, H.G. (1979). The nutrient budgets of even-aged forests. In *Ecology of even-aged forest plantations* (ed. E.D. Ford, D.C. Malcolm, and J. Atterson), pp. 221–56. Institute of Terrestrial Ecology, Cambridge.

Miller, H.G. (1981*a*). Nutrient cycles in forest plantations, their change with age and the consequence for fertilizer practice. In *Proceedings of Australian forest nutrition workshop: productivity in perpetuity*, pp. 187–99. Canberra, Australia, 10–14 August 1981. CSIRO Division of Forest Research.

Miller, H.G. (1981*b*). Forest fertilization: some guiding concepts. *Forestry*, **54**, 157–67.

Miller, H.G. (1981*c*). Aspects of forest fertilization practice and research in New Zealand. *Scottish Forestry*, **35**, 277–88.

Miller, H.G. (1984). Nutrition of hardwoods. In *Report of fifth meeting of National Hardwoods Programme*, pp. 17–29. Commonwealth Forestry Institute, Oxford.

Miller, H.G. (1995). The influence of stand development on nutrient demand, growth and allocation. In *Nutrient uptake and cycling in forest ecosystems* (ed. R.F. Hüttl, L.O. Nilsson, and U.T. Johansson). *Plant and Soil,* pp. 168–9.

Miller, H.G., Williams, B.L., Millar, C.S., and Warin, T.R. (1977). Ground vegetation and humus nitrogen levels as indicators of nitrogen status in an established sand dune forest. *Forestry*, **50**, 93–101.

Miller, H.G., Miller, J.D., and Cooper, J.M. (1981). Optimum foliar nitrogen concentration in pine and its change with stand age. *Canadian Journal of Forest Research*, **11**, 563–72.

Miller, J.F. (1985). *Windthrow hazard classification*. Forestry Commission Leaflet, 85. HMSO, London.

Miller, K.F., Quine, C.P., and Hunt, J. (1987). The assessment of wind exposure for forestry in upland Britain. *Forestry*, **60**, 179–92.

Mills, D.H. (1980). *The management of forest streams*. Forestry Commission Leaflet, 78. HMSO, London.

Mitchell, A.F. (1988). *A field guide to the trees of Britain and northern Europe*. Collins, London.

Mitchell, C.P., Ford-Robertson, J.B., Hinckley, T., and Sennerby–Forsse, L. (ed.) (1992). *Ecophysiology of short rotation forest crops*. Elsevier, London.

Mitscherlich, E.A. (1921). Das wirkungsgesetz der Wachstumfaktoren. *Landwirtschaft Jahrbuch Bog*, 11–5.

Mitscherlich, G. (1973). [Forest and wind.] *Allgemeine Forst– und Jagdzeitung*, **144**, 76–81.

Moffat, A.J. (1987). The geological input to the reclamation process in forestry. In *Planning and engineering geology* (ed. M.G. Culshaw, F.G. Bell, J.C. Cripps, and M. O'Hara), pp. 541–8. Engineering Geology Special Publication, **4**, Geological Society, London.

Moffat, A.J. and McNeill, J.D. (1994). *Reclaiming disturbed land for forestry*. Forestry Commission Bulletin, 110. HMSO, London.

Moffat, A.J. and Roberts, C.J. (1989). Use of large-scale ridge and furrow landforms in forestry reclamation of mineral workings. *Forestry*, **62**, 233–48.

Møller, C.M. (1933). *Boniteringstabeller og bonitetsvise tillväxtoversikter for Bog, Eg og Rödgran i Danmark*. Dansk Skogforenings Tidskrift No. 18.

Møller, C.M. (1960). The influence of pruning on the growth of conifers. *Forestry*, **33**, 37–53.

Molnar, A. (1989). *Community forestry—a review*. Community Forestry Note 3. FAO, Rome.

Monaco, L.C. (1983). Bioenergy in the north-south dialogue. In *Energy from Biomass* (ed. A. Strub, P. Chartier, and G. Schleser), pp. 36–42. Applied Science Publishers, London.

Moore, P.D. (1984). Why be an evergreen? *Nature*, **312**, 703.

Moran, G.F. and Bell, J.C. (1987). The origin and genetic diversity of *Pinus radiata* in Australia. *Theoretical and Applied Genetics*, **73**, 616–22.

Moreno, J.M. and Oechel, W.C. (ed.) (1994). *The role of fire in Mediterranean–type ecosystems*. Ecological studies, 107. Springer-Verlag, New York.

Moss, D. (1979). Even-aged plantations as a habitat for birds. In *Ecology of even-aged forest plantations* (ed. E.D. Ford, D.C. Malcolm, and J. Atterson), pp. 413–27. Institute of Terrestrial Ecology, Cambridge.

Müller-Starck, G. and Ziehe, M. (ed.) (1991). *Genetic variation in European populations of forest trees*. J.D. Sauerland's Verlag, Frankfurt am Maine, Germany.

Mullin, R.E. (1974). Some planting effects still significant after 20 years. *Forestry Chronicle*, **50**, 191–3.

Murray, J.S. (1979). The development of populations of pests and pathogens in even-aged plantations—fungi. In *Ecology of even-aged forest plantations* (ed.

E.D. Ford, D.C. Malcolm, and J. Atterson), pp. 193–208. Institute of Terrestrial Ecology, Cambridge.

Nair, P.K.R. (1993). *An introduction to agroforestry*. Kluwer Academic, London.

Nambiar, E.J.S. (1981). Ecological and physiological aspects of the development of roots: from nursery to forest. In *Proceedings of the Australian forest nutrition workshop*, Canberra, ACT, Australia, pp. 117–29.

Namkoong, G. (1989). Systems of gene management. In *Breeding tropical trees* (ed. G.L. Gibson, A.R. Griffin, and A.C. Matheson), pp. 1–8. Oxford Forestry Institute and Winrock International.

National Research Council (USA) (1991). *Managing global genetic resources— tropical forest trees*. Commonwealth Forestry Institute, University of Oxford, (Washington, DC: Tropical Forestry Papers No. 16. National Academy Press).

Naveh, Z. (1975). The evolutionary significance of fire in the Mediterranean region. *Vegetatio*, **29**, 199–208.

Neckelmann, J. (1981). [Stabilization of edges and internal shelter zones in stands of Norway spruce on sandy soil.] *Dansk Skovforenings Tidsskrift*, **66**, 196–314.

Neckelmann, J. (1982). [Stabilizing measures in Norway spruce—and the hurricane of November 1981.] *Dansk Skovforenings Tidsskrift*, **67**, 77–86.

Nelson, D.G. and Quine, C.P. (1990). *Site preparation for restocking*. Forestry Commission Research Information Note, 166. Farnham, Surrey, UK.

Nepveu, G. and Blachon, J.L. (1989). Largeur de cerne et aptitude à l'usage en structure de quelques conifères: Douglas, pin sylvestre, pin maritime, épicéa de Sitka, épicéa commun, sapin pectiné. *Revue Forestière Française*, **41**, 497–506.

Nepveu, G. and Velling, P. (1983). Variabilité génétique individuelle de la qualité du bois de *Betula pendula*. *Silvae Genetica*, **32**, 37–49.

Nepveu, G., Bailly, A., and Coquet, M. (1985). [The susceptibility of *Picea excelsa* to wind damage may be explained by the wood density being too low.] *Revue Forestière Française*, **37**, 305–8.

Neustein, S.A. (1964). Windthrow on the margins of various sizes of felling area. In *Forestry Commission Report on Forest Research 1964*, pp. 166–71. HMSO, London.

Newman, S.M. (1994). An outline comparison of approaches to silvo-arable research and development with fast-growing trees in India, China, and the UK with emphasis on intercropping with wheat. *Agroforestry Forum*, **5**, 29–31.

Newnham, R.M. (1965). Stem form and the variation of taper with age and thinning regime. *Forestry*, **38**, 218–24.

Norgren, O. (1995). Growth differences between *Pinus sylvestris* and *Pinus contorta*.Dissertation, Department of Silviculture, Swedish University of Agricultural Sciences, Umeå.

O'Carroll, N. (1978). The nursing of Sitka spruce: 1. Japanese larch. *Irish Forestry*, **35**, 60–5.

O'Carroll, N., Carey, M.L., Hendrick, E., and Dillon, J. (1981). The tunnel plough in peatland afforestation. *Irish Forestry*, **38**, 27–40.

O'Driscoll, J. (1980). The importance of lodgepole pine in Irish forestry. *Irish Forestry*, **37**, 7–22.

Office National des Forêts (1989). *Manuel d'aménagement*. Office National des Forêts, Paris.

Ogilvie, J.F. and Taylor, C.S. (1984). Chemical silviculture (chemical thinning and respacement). *Scottish Forestry*, **38**, 83–8.

Olesen, F. (1979). [*Planting Shelterbelts.*] Laeplanting, Denmark. Landhusholdnings selskabet, Copenhagen.

Olesen, F. (1993). Investigations of shelter effect and experience with wind protection in Danish agriculture. In *Windbreaks and agroforestry*, pp. 210–13. Hedeselskabet, Viborg, Denmark.

Oliveira, M.A. de and Oliveira, L. de (1991). *Le liège*. Amorim, Spain.

Oliver, C.D. and Larson, B.C. (1996). *Forest stand dynamics*. Wiley, New York.

Örlander, G., Gemmel, P., Hunt, J. (1990). *Site preparation: a Swedish overview*. British Columbia Ministry of Forestry, Canada.

Örlander, G., Halsby, G., Gemmel, P., and Wilhelmsson, C. Inverting site preparation improves survival and growth of lodgepole pine and Norway spruce. *Canadian Journal of Forest Research* (in press).

Oswald, H. (1980). Une expérience d'espacement de plantation de Sapin de Vancouver (*Abies grandis*). *Revue Forestière Française*, **32**, 60–77.

Oswald, H. and Pardé, J. (1976). Une expérience d'espacement de plantation de Douglas en forêt domaniale d'Amance. *Revue Forestière Française*, **28**, 185–92.

Otto, H.-J. (1976). [Forestry experience and conclusions from the forest catastrophes in Lower Saxony.] *Forst und Holzwirt* **15**, 285–95. (British Lending Library *Translation RTS 1* 1890).

Otto, H.-J. (1982). Measures to reduce forest fire hazards and restoration of damaged trees in Lower Saxony. In *Fire prevention and control* (ed. T. van Nao), pp. 173–9. Martinus Nijhoff/W Junk, The Hague.

Paavilainen, E. and Päivänen, J. (1995). *Peatland forestry: ecology and principles*. Springer–Verlag, Berlin Heidelberg.

Papesch, A.J.G. (1974). A simplified theoretical analysis of the factors that influence windthrow of trees. In *5th Australasian Conference on Hydraulics and Fluid mechanics*, pp. 235–42. University of Canterbury, New Zealand.

Pardé, J. (1980). Forest biomass. *Forestry Abstracts* (Review Article), **41**, 343–62.

Pardé, J. and Bouchon, F. J. (1988). *Dendrométrie*. ENGREF, Nancy, France.

Parsons, A.D. and Evans, J. (1977). Forest fire protection in the Neath district of south Wales. *Quarterly Journal of Forestry*, **71**, 186–98.

Pawsey, C.J. (1972). Survival and early development of *Pinus radiata* as influenced by size of planting stock. *Australian Forestry Research*, **5**, 13–24.

Pepper, H.W. (1978). *Chemical repellents*. Forestry Commission Leaflet, 73. HMSO, London.

Pereira, H. (1995). Silvicultural management of cork–oak stands towards improved cork production and quality. In *Non-food, bio-energy and forestry* (ed. C. Mangan, B. Kerckow, and M. Flanagan), pp. 286–7. European Commission EUR 16206EN, Brussels.

Pereira, H. and Santos Pereira, J.(1988). Short rotation biomass plantations in Portugal. In *Biomass forestry in Europe: a strategy for the future* (ed. F.C. Hummel, W. Palz, and G. Grassi), pp. 509–39. Elsevier, London.

Perry, D.A. (1979). Variation between and within tree species. In *Ecology of even-aged forest plantations* (ed. E.D. Ford, D.C. Malcolm, J. Atterson), pp. 71–98. Institute of Terrestrial Ecology, Cambridge.

Persson, A. (1980). *Pinus contorta as an exotic species.* Department of Forest Genetics, Swedish University of Agricultural Sciences, Garpenberg. Research Notes, 30, p.15.

Persson, O.A. (1992). [*A growth simulator for Scots pine in Sweden*]. Department of Forest Yield Reseach, Swedish University of Agricultural Sciences, Garpenberg. Report No. **31**.

Persson, P. (1975). [*Windthrow in forests: its cause and the effect of forestry measures.*] Rapporter och Uppsater, Institutionen för Skogsproduktion No. 36.

Peterken, G.F. (1981). *Woodland conservation and management.* Chapman and Hall, London.

Peterken, G.F. (1996). *Natural woodland.* Cambridge University Press.

Pettersson, F. (1994). *Predictive functions for calculating the total response in growth to nitrogen fertilization, duration and distribution over time.* Skog Forsk Report No. 4. Oskarshamn, Sweden.

Petty, J.A. and Worrell, R. (1981). Stability of coniferous tree stems in relation to damage by snow. *Forestry*, **54**, 115–28.

Philipson, J.J. and Coutts, M.P. (1978). The tolerance of tree roots to waterlogging: III. Oxygen transport in lodgepole pine and Sitka spruce roots of primary structure. *New Phytologist*, **80**, 341–9.

Philipson, J.J. and Coutts, M.P. (1980). The tolerance of tree roots to waterlogging. IV. Oxygen transport in woody roots of Sitka spruce and lodgepole pine. *New Phytologist*, **85**, 489–94.

Poore, D. and Sayer, J. (1987). *The management of tropical moist forest lands: ecological guidelines.* International Union for the Conservation of Nature , Gland, Switzerland.

Potter, M. (1991). *Treeshelters.* Forestry Commission Handbook, 7. HMSO, London.

Price, C. (1989). *The theory and application of forest economics.* Basil Blackwell, Oxford.

Price, C. (1993). *Time, discounting and value.* Basil Blackwell, Oxford.

Prior, R. (1983). *Trees and deer.* B.T. Batsford, London.

Pritchett, W.L. (1979). *Properties and management of forest soils.* Wiley, New York.

Pyatt, D.G. (1990). *Forest drainage schemes.* Forestry Commission Research Information Note, 196. Farnham, Surrey, UK.

Pyatt, D.G. and Craven, M.M. (1979). Soil changes under even-aged plantations. In *Ecology of even-aged forest plantations* (ed. E.D. Ford, D.C. Malcolm, and J. Atterson), pp. 369–86. Institute of Terrestrial Ecology, Cambridge.

Pyne, S.J. (1984). *Introduction to wildland fire: fire management in the United States*. Wiley, New York.

Quimby, P.C. (1982). Impact of diseases on plant populations. In *Biological control of weeds with plant pathogens* (ed. R. Charudattan and H.L. Walker), pp. 47–60. Wiley, New York.

Quine, C.P. (1989). Description of the storm and comparison with other storms. In *Forestry Commission Bulletin*, 87, pp. 3–8. HMSO, London.

Quine, C.P. (1992). Windthrow as a constraint on silviculture. In *Whither silviculture*. Proceedings of a symposium 28–29 November 1991, pp. 21–9. Institute of Chartered Foresters, Edinburgh.

Quine, C.P. and Miller, K.F. (1990). Windthrow—a factor influencing the choice of silvicultural system. In *Silvicultural systems* (ed. P. Gordon), pp. 71–81. Institute of Chartered Foresters, Edinburgh.

Quine, C.P. and White, I.M.S. (1993). *Revised windiness scores for the windthrow hazard classification: the revised scoring method*. Forestry Commission Research Information Note, 230, Farnham, UK.

Quine, C.P., Burnand, A.C., Coutts, M.P., and Reynard, B.R. (1991). Effects of mounds and stumps on the root architecture of Sitka spruce on a peaty gley restocking site. *Forestry*, **64**, 385–401.

Quine, C.P., Coutts, M.P., Gardiner, B.A., and Pyatt, D.G. (1995). *Forests and wind: management to minimize damage*. Forestry Commission Bulletin, 114. HMSO, London.

Rackham, O. (1976). *Trees and woodland in the British landscape*. J.M. Dent and Sons, London.

Raintree, J.B. (1987). The state of the art of agroforestry diagnosis and design. *Agroforestry Systems*, **5**, 219–50.

Randall, A. (1987). *Resource economics*, (2nd edn.). Wiley, New York.

Ranger, J. and Nys, C. (1996). Biomass and nutrient content of extensively and intensively managed coppice stands. *Forestry*, **69**, 83–102.

Ranger, J., Barnéoud, C., and Nys, C. (1988). Production ligneuse et rétention d'éléments nutritifs par des taillis à courte rotation de peuplier 'Rochester': effet de la densité d'ensouchement. *Acta Œcologica*, **9**, 245–69.

Ranger, J., Nys, C., and Barnéoud, C. (1986). Production et exportation d'éléments nutritifs de taillis de peuplier à courte rotation. In *Annales AFOCEL* 1986, pp. 183–223.

Ranger, J., Robert, M., Bonnaud, P., and Nys, C. (1991). Les minéraux-tests: une approche expérimentale *in situ* de l'altération biologique et du fonctionnement des écosystèmes forestiers. Effets des types de sols et des essences feuillues et résineuses. *Annales des Sciences Forestières*, **47**, 529–50.

Rapey, H. (1994). Les vergers à bois précieux en prairie pâturée: objectifs, principes et références. *Revue Forestière Française*, **46**, 61–71.

Rapey, H., Montard, F.X. de, and Guitton, J.L. (1994). Ouverture de plantations résineuses au pâturage: implantation et production d'herbe dans le sous-bois après éclaircie. *Revue Forestière Française*, **46**, 19–29.

Ratcliffe, P.R. (1987). *The management of red deer in upland forests*. Forestry Commission Bulletin, 71. HMSO, London.

Ratcliffe, P.R. and Mayle, B.A. (1992). *Roe deer biology and management*. Forestry Commission Bulletin, 105. HMSO, London.

Raymer, W.G. (1962). *Wind resistance of conifers*. Report, 1008. National Physical Laboratory Aerodynamics Division, UK.

Rees, D.J. and Grace, J. (1980*a*). The effects of wind on the extension growth of *Pinus contorta*. *Forestry*, **53**, 145–53.

Rees, D.J. and Grace, J. (1980*b*). The effects of shaking on extension growth of *Pinus contorta*. *Forestry*, **53**, 155–66.

Reineke, L.H. (1933). Perfecting a stand-density index for even-aged forests. *Journal of Agricultural Research,* **46**, 627–38.

Richter, J. (1975). [Gale damage to spruce in Sauerland.] *Forst– und Holzwirt,* **30**, 106–8.

Rickman, R. (1991). *What's good for woods*. Policy Study No. 129. Centre for policy studies, London.

RICS (Royal Institution of Chartered Surveyors) (1996). *Lowland forestry on traditional estates*. Royal Institution of Chartered Surveyors, London.

Rigolot, E. (1993). Le brûlage dirigé en région méditerranéenne française. In *Rencontres forestiers—chercheurs en forêt méditerranéenne*, les Colloques No. 63, (ed. H. Oswald), pp. 223–50. INRA, Paris.

Ritchie, G.A. and Dunlap, J.R. (1980). Root growth potential: its development and expression in forest tree seedlings. *New Zealand Journal of Forestry Science*, **10**, 218–48.

Rodwell, J. and Patterson, G. (1994). *Creating new native woodlands*. Forestry Commission Bulletin, 112. HMSO, London.

Rollinson, T.J.D. (1983). Mensuration. In *Forestry Commission Report on Forest Research1983 , pp. 44–45*. HMSO, London.

Rosenberg, N.J. (1974). *Microclimate: the biological environment*. Wiley, New York.

Ross, S.M. and Malcolm, D.C. (1982). Effects of intensive forestry ploughing practices on an upland heath soil in south-east Scotland. *Forestry*, **55**, 155–71.

Rothermel, R.C. (1982). Modelling the development of fire in a forest environment. In *Forest fire prevention and control* (ed. T. van Nao), pp. 77–84. Martinus Nijhoff/W Junk, The Hague, Netherlands.

Rowe, J.S. (1983). Concepts of fire effects on plant individuals and species. In *The role of fire in northern circumpolar ecosystems* (ed. R.W. Wein and D.A. MacLean), pp. 135–54. Wiley, New York.

Sanders, F.E. and Tinker, P.B. (1973). Phosphate flow in mycorrhizal roots. *Pesticide Science*, **4**, 385–95.

Sanderson, P.L. and Armstrong, W. (1978). Soil waterlogging, root rot and conifer windthrow: oxygen deficiency or phytotoxicity? *Plant and Soil*, **49**, 185–90.

Saracino, A. and Leone, V. (1994). The ecological role of fire in Aleppo pine forests: overview of recent research. In *Forest fire research*, 2nd International Conference, 2, pp. 887–97. University of Coimbra, Portugal.

Sargent, C. (1990). *The Khun Song Plantation Project*. International Institute for Environment and Development, London.

Sargent, C. and Bass, S. (1992). *Plantation politics*. Earthscan Publications, London.

Saur, E. (1993). Interactive effects of P–Cu fertilizers on growth and mineral nutrition of maritime pine. *New Forests*, **7**, 93–105.

Savill, P.S. (1976). The effects of drainage and ploughing of surface water gleys on rooting and windthrow of Sitka spruce in Northern Ireland. *Forestry*, **49**, 133–41.

Savill, P.S. (1983). Silviculture in windy climates. *Forestry Abstracts* (Review Article), **44**, 473–88.

Savill, P.S. (1991). *The silviculture of trees used in British forestry*. CAB International, Wallingford, UK.

Savill, P.S. and Mather, R.A. (1990). A possible indicator of shake in oak: relationship between flushing dates and vessel sizes. *Forestry*, **63**, 355–62.

Savill, P.S. and McEwen, J.E. (1978). Timber production from Northern Ireland 1980–2004. *Irish Forestry*, **35**, 115–23.

Savill, P.S. and Sandels, A.J. (1983). The influence of spacing on the wood density of Sitka spruce. *Forestry*, **56**, 109–20.

Savill, P.S. and Spilsbury, M.J. (1991). Growing oaks at closer spacing. *Forestry*, **64**, 373–84.

Savill, P.S., Dickson, D.A., and Wilson, W.T. (1974). Effects of ploughing and drainage on growth and root development of Sitka spruce on deep peat in Northern Ireland. In *Proceedings International Union of Forestry Research Organisations Symposium on Forest Drainage*, Helsinki.

Saxena, N.C. (1991). Marketing constraints for *Eucalyptus* from farm lands in India. *Agroforestry Systems*, **13**, 73–86.

Sayer, J.A. and Whitmore, T.C. (1991). Tropical moist forests: destruction and species extinction. *Biological Conservation*, **55**, 199–213.

Schlaepfer, R. (ed.) (1993). *Long-term implications of climatic change and air pollution on forest ecosystems*. Progress report of the International Union of Forestry Research Organisations task force 'Forest, climate change and air pollution'. International Union of Forestry Research Organisations World Series, Vol. 4. International Union of Forestry Research Organisations Secretariat, Vienna.

Schlich, W. (1899). *Manual of forestry*, Vol. I. Bradbury, Agnew and Co., London.

Schmutz, T. (1994). Quinze ans de replantations en France. *Revue Forestière Française*, **46**, 119–24.

Schoenweiss, D.F. (1981). The role of environmental stress in disease of woody plants. *Plant Diseases*, **65**, 308–14.

Schulze, E.-D. (1982). Plant life forms and their carbon, water and nutrient relations. In *Physiological plant ecology* II (ed. O.L. Lange, P.S. Nobel, C.B. Osmond, and H. Ziegler), pp. 615–76. Springer-Verlag, Berlin.

Schwappach, A. (1902). *Waschstum und Ertrag normaler Fichtenbestände in Preussen. Mitteilungen aus dem forstlichen Versuchswesen Preussens.* Neudamm.

Seligman, R.M. (1983). Biofuels in the European Community—a view from the European Parliament. In *Energy from biomass* (ed. A. Strub, P. Chartier, and G. Schleser), pp. 16–22. Applied Science Publishers, London.

Sendak, P.E. (1978). Birch sap utilization in the Ukraine. *Journal of Forestry*, **76**, 120–21.

Sennerby–Forsse, L. and Johansson, H. (ed.) (1989). *Handbook for energy forestry*. Swedish University of Agricultural Sciences, Uppsala, Sweden.

Seuna, P. (1981). *Long-term influence of forestry drainage on the hydrology of an open bog in Finland*. Publication, 43, pp. 3–14. Water Research Institute, Finland.

Sheehan, P.G., Lavery, P.B., and Walsh, B.M. (1982). Thinning and salvage strategies in plantations prone to storm damage. *New Zealand Journal of Forestry Science*, **12**, 169–80.

Sheldrick, R. and Auclair, D. (1995). The development of non–tropical agroforestry systems. *Agroforestry Forum,* **6**, 58–61.

Shell/WWF (1993). *Tree plantation review*. Shell International Petroleum Co. Ltd. and World Wide Fund for Nature.

Shellard, H.C. (1976). Wind. In *The climate of the British Isles* (ed. T.J. Chandler and S. Gregory), pp. 39–73. Longman, London.

Shin, D.I., Podila, G.K., Huang, Y., Karnosky, D.F., and Huang, Y.H. (1994). Transgenic larch expressing genes for herbicide and insect resistance. *Canadian Journal of Forest Research*, **24**, 2059–67.

Shirley, H.L. (1945). Reproduction of upland conifers in the Lake States as affected by root competition and height. *American Midland Naturalist*, **33**, 537–611.

Simpson, J. (1900). *The new forestry*. Pawson and Brailsford, Sheffield.

Skidmore, E.L. (1976). Barrier–induced microclimate and its influence on growth and yield of winter wheat. In *Shelterbelts on the Great Plains* (ed. R.W. Tinus). Great Plains Agricultural Publication, 78, pp. 57–63.

Slodicak, M. (1987). Resistance of young spruce stands to snow and wind damage in dependence on thinning. *Communicationes Instituti Forestalis Cechosloveniae*, **15**, 75–86.

Smith, H.C. and Gibbs, C.B. (1970). *A guide to sugarbush stocking*. USDA Forest Service Research Paper NE–171.

Söderberg, U. (1986). [*Functions for forecasting of timber yields*]. Report No. 14 Department of Measuration and Management, Swedish University of Agricultural Sciences, Umeå.

Söderström, V. (1976). Markvärme–en minimifaktor vid plantering [Soil temperature: a minimum factor when planting]. *Skogsarbeten, Redogörelse*, **6**,16–22.

Sol, B. (1990). Estimation du risque météorologique d'incendies de forêts dans le sud–est de la France. *Revue Forestière Française*, **42**, No. sp, 263–71.

Southwood, T.R.E. (1981). Bionomic strategies and population parameters. In *Theoretical ecology* (2nd edn) (ed. R.M. May), pp. 30–52. Blackwell, Oxford.

Speight, M.R. (1983). The potential of ecosystem management for pest control. *Agriculture, Ecosystems and Environment*, **10**, 183–99.

Speight, M.R. and Wainhouse, D. (1989). *Ecology and management of forest insects*. Clarendon Press, Oxford.

Spilsbury, M.J. (1990). *Modelling the development of mixed deciduous woodland ecosystems*. Department of Plant Sciences, University of Oxford, unpublished DPhil thesis.

Stassen, H.E.M. (1982). [Energy from wood and wood waste; technologies and perspectives.] *Nederlands Bosbouw Tijdschrift*, **54**, 172–8.

Steele, R.C. (1972). *Wildlife conservation in woodlands*. Forestry Commission Booklet, 29. HMSO, London.

Stein, W.I. (1978). Naturally developed seedling roots of five western conifers. In *Proceedings of the root form of planted trees symposium* (ed. E. van Eerden and J.M. Kinghorn), pp. 28–35. British Columbia Ministry of Forests/Canadian Forestry Service Joint Report 8.

Stéphan, J.M. (ed.) (1994). *Feux de forêt: bilans 93–94*. Ministère de l'Agriculture, Ministère de l'Intérieur, DERF, Paris.

Stewart, A.J.A. and Lance, A.N. (1983). Moor-draining: a review of impacts on land use. *Journal of Environmental Management*, **17**, 81–99.

Stewart, P.J. (1987). *Growing against the grain*. Council for the Protection of Rural England, Oxford.

Stocker, G.C. (1976). *Report on cyclone damage to natural vegetation in the Darwin area after cyclone Tracey, 25 December 1974*. Forestry and Timber Bureau Leaflet, 127. Canberra.

Streets, R.J. (1962). *Exotic forest trees in the British Commonwealth*. Clarendon Press, Oxford.

Sutton, R.F. (1993). Mounding in site preparation. A review of European and north American experience. *New Forests*, **7**, 151–92.

Sutton, W.R.J. (ed.) (1970). Pruning and thinning practice. In *Proceedings of New Zealand Forest Service, Forest Research Institute Symposium*, Vol. 2.

Sutton, W.R.J. (1984). Economic and strategic implications of fast-growing plantations. In *International Union of Forestry Research Organisations Symposium on site and productivity of fast-growing plantations*, Vol. 1, pp. 417–31. Forest Research Institute, Pretoria, South Africa.

Sweet, G.B. and Waring, P.F. (1966). The relative growth rate of large and small seedlings in forest tree species. *Forestry* (supplement), **39**, 110–7.

Symonds, H.H. (1936). *Afforestation in the Lake District*. J.M. Dent and Sons, London.

Tabbush, P.M. (1987). Effect of desiccation on water status and forest performance of bare-rooted Sitka spruce and Douglas fir transplants. *Forestry*, **60**, 31–43.

Tabbush, P.M. (1988). *Silvicultural principles for upland restocking*. Forestry Commission Bulletin, 76. HMSO, London.

Tabbush, P.M. and Williamson, D.R. (1987). *Rhododendron ponticum* as a forest weed. Forestry Commission Bulletin, 73. HMSO, London.

Tadaki, Y. (1966). Some discussions on the leaf biomass of forest stands and trees. *Bulletin of the Government Forest Experiment Station, Meguro*, **84**, pp. 135–61.

Talamucci, P. (1989). Choix des espèces ligneuses et leur production fourragère en Italie. In *Les espèces ligneuses à usages multiples des zones arides méditerranéennes* (ed. R. Morandini), pp. 40–58. Commission of the European Communities report EUR 11770, Luxemburg.

Taylor, C.M.A. (1991). *Forest fertilization in Britain*. Forestry Commission Bulletin, 95. HMSO, London.

Taylor, G.G.M. (1970). *Ploughing practice in the Forestry Commission*. Forestry Commission Forest Record, 73. HMSO, London.

Taylor, J.A. (1976). Upland climates. In *The climate of the British Isles* (ed. T.J. Chandler and S. Gregory), pp. 264–87. Longman, London.

Templeton, G.E. (1981). Status of weed control with plant pathogens. In *Biological control of weeds with plant pathogens* (ed. R. Charudattan and H.L. Walker), pp. 29–44. Wiley, New York.

Tessier du Cros, É. (1994). Génétique et amélioration des arbres forestiers. *Forêt Entreprise*, **96**, 15–16.

Thomas, J.W., Miller, R.J., Black, H., Rodiek, J.E., and Maser, C. (1976). Guidelines for maintaining and enhancing wildlife habitat in forest management in the Blue Mountains of Washington and Oregon. *Transactions of the north American Wildlife and Natural Resources Conference*, **41**, 452–76.

Thompson, D.A. (1979). *Forest drainage schemes*. Forestry Commission Leaflet, 72. HMSO, London.

Thompson, D.A. (1984). *Ploughing of forest soils*. Forestry Commission Leaflet, 71. HMSO, London.

Toleman, R.D.L. and Pyatt, D.G. (1974). Site classification as an aid to silviculture in the Forestry Commission of Great Britain. *Paper for 10th Commonwealth Forestry Conference*, UK, 1974.

Tombleson, J. (1993). Timber production from shelterbelts—The New Zealand experience. In *Windbreaks and agroforestry*, pp. 39–43. Hedeselskabet, Viborg, Denmark.

Tombleson, J. and Inglis, C.S. (1988). Comparison of radiata pine shelterbelts and plantations. In *Agroforestry symposium proceedings* (ed. P. MacLaren), pp. 261–78. New Zealand Ministry of Forestry, Forest Research Institute, Bulletin, 139.

Trabaud, L. (1981). Man and fire: impacts on Mediterranean vegetation. In *Mediterranean–type shrublands* (ed. F. di Castri, D.W. Goodall, and R.L. Specht), pp. 523–37. Elsevier, Amsterdam.

Trabaud, L. (1989). *Les feux de forêt—mécanismes, comportements et environnement*. France sélection, Aubervilliers, France.

Tranquillini, W. (1979). *Alpine timberline*. Springer-Verlag, Berlin.

Tribun, P.A., Gavrilyuk, M.V., Yukhimchuk, G.V., and Lopareva, E.B. (1983). [Biochemical features of young spruce trees in stands of different density.] *Lesnoi Zhurnal*, **3**, 23–6.

Troedsson, T. (1980). Long-term changes of forest soils. *Annales Agriculturae Fenniae*, **19**, 81–4.

Tuley, G. (1983). Shelters improve the growth of young trees. *Quarterly Journal of Forestry*, **77**, 78–87.

Tuley, G. (1985). The growth of young oak trees in shelters. *Forestry*, **58**, 181–95.

UN–ECE/FAO (United Nations–ECE/Food and Agriculture Organization) (1992). *The forest resources of the temperate zone*. United Nations, New York.

Upton, C. and Bass, S. (1995). *The forest certification handbook*. Earthscan Publications, London.

Valadon, A. (1996). Évolution de la populiculture, période 1992–1995. Rapport national de la France. In *International poplar commission, 20th meeting, Hungary*.

Valette, J.C. (1990). Inflammabilités des espèces forestières méditerranéennes. Conséquences sur la combustibilité des formations forestières. *Revue Forestière Française*, **42**, No. sp., 76–92.

Vannière, B. (ed) (1984). *Tables de production pour les forêts françaises*. Ecole Nationale du Génie Rural et des Eaux et Forêts, Nancy, France. (2nd edn).

Vincent, J.R. and Binkley, C.S. (1992). Forest-based industrialization:a dynamic perspective. In *Managing the world's forests* (ed. N.P. Sharma). Kendall/Hunt Publishing Company, Iowa, 93–137.

Viro, P.J. (1969). *Prescribed burning in forestry*. Communicationes Instituti Forestalis Fenniae.

Vuokila, Y. and Valiaho, H. (1980). [*Growth and yield models for conifer cultures in Finland*]. Communicationes Instituti Forestale Fenniae, 99 (2).

Walshe, D.E. and Fraser, A.I. (1963). *Wind tunnel tests on a model forest*. Report 1078, National Physical Laboratory Aerodynamics Division, UK.

Wang, T.-T., Pai, N.-Y., Peng, P.L.F., Lin, T.-S., and Shih, C.-F. (1980). The effect of pruning on the growth of *Cryptomeria*. Biologische, technische und wirtschaftliche Aspekte der Jungbestand ege (ed. H. Kramer). *Schriften Forstlichen Fakultät*, **67**, 92–106. Universität Göttingen, Germany.

Watts, S.B. (1983). *Forestry handbook for British Columbia*. University of British Columbia, Canada.

Webber, J.F. and Gibbs, J.N. (ed.) (1996). *Water storage of timber: experience in Britain*. Forestry Commission Bulletin, 117. HMSO, London.

Wein, R.W. and MacLean, D.A. (ed.) (1983). *The role of fire in northern circumpolar ecosystems.* Wiley, New York.

Westoby, J.C. (1962). The role of forest industries in the attack on economic underdevelopment. In: Westoby, J.C. (1987). *The purpose of forests.* Basil Blackwell, Oxford, pp. 3–70.

Westoby, J.C. (1987). In *The purpose of forests.* Basil Blackwell, Oxford.

White, T.L. (1987). A conceptual framework for tree improvement programs. *New Forests,* **1**, 325–42.

Whitehead, D. (1981). Ecological aspects of natural and plantation forests. *Forestry Abstracts,* (Review article), **43**, 615–24.

Whiteside, I.D. (1989). Economic advances made in New Zealand radiata pine plantation forestry since the early 1980s. In *Proceedings 13th Commonwealth Forestry Conference,* **6C**. Rotorua, New Zealand.

Whyte, A.G.D. (1988). Radiata pine silviculture in New Zealand: its evolution and future prospects. *Australian Forestry,* **51**, 185–96.

Wickens, D., Rumfitt, A., and Willis, R. (1995). *Survey of derelict land in England 1993.* Department of the Environment, HMSO, London.

Wiedermann, E. (1923). [*Regress in increment and growth interruptions of spruce in middle and lower altitudes of the Saxon State forests.*] Translation No. 301, United States Forest Service, 1936.

Wiedermann, E. (1937). *Die Fichte.* Mitteilungen aus Forstwirtschaft und Forstwissenschaft, 1936.

Wiedermann, E. (1948). *Die Kiefer. Waldbauliche und ertragskundliche Untersuchungen.* Verlag M & H Schaper, Hannover.

Wiedermann, E. (1949). *Ertragstafeln der wichtigen Holzarten.* Verlag M & H Schaper, Hannover.

Williamson, D.R. (1990). The use of herbicides in UK forestry. In *Tendencias mundialesen el control de la vegetacion accesoria en los montes.* Universidad politecnica de Madrid and Asociacion ingenieros montes, Madrid.

Willoughby, I. and Dewar, J. (1995). *The use of herbicides in the forest.* Forestry Commission Field Book, 8. HMSO, London.

Wilson, C.L. (1969). Use of plant pathogens in weed control. *Annnal Review of Phytopathology,* **7**, 411–34.

Wilson, K. and Pyatt, D.G. (1984). An experiment in intensive cultivation of an upland heath. *Forestry,* **57**, 117–41.

Wilson, R.V. (1989). Financial returns from plantation forestry in Australia. In *Proceedings 13th Commonwealth Forestry Conference,* 6C. Rotorua, New Zealand.

Winer, N. (1980). The potential of the carob. *International Tree Crops Journal,* **1**, 15–26.

Winpenny, J.T. (1991). *Values for the environment.* HMSO, London.

Wolstenholme, R., Dutch, J., Moffat, A.J., Bayes, C.D., and Taylor, C.M.A. (1992). *A manual of good practice for the use of sewage sludge in forestry.* Forestry Commission Bulletin, 107. HMSO, London.

Wright, H.A. and Bailey, A.W. (1982). *Fire ecology*. Wiley, New York.

Wright, L.L. (1988). Are increased yields in coppice systems a myth? *Finnish Forest Research Institute Bulletin*, **304**, 51–65.

Yoda, K., Kira, T., Ogawa, H., and Hozumi, J. (1963). Self-thinning in overcrowded pure stands under cultivated and natural conditions. *Journal of Biology Osaka City University*, **14**, 107–29.

Yoho, J.G. (1985). Continuing investments in forestry: private investment strategies. In *Investment in forestry* (ed. R.A. Sedjo). Westview Press, Boulder, CO, USA.

Ziller, W.G. (1967). *The tree rusts of Western Canada*. Canadian Forest Service Publication, 1329. Department of the Enviroment, Victoria.

Zobel, B.J. and Talbert, J. (1984). *Applied forest tree improvement*. Wiley, New York.

Zobel, B.J., Campinhos, E., and lkemori, Y. (1983). Selecting and breeding for desirable wood. *Tappi*, 70–4.

Zobel, B.J., Wyk, G. van, and Stahl, P. (1987). *Growing exotic forests*. Wiley, New York.

Index

Tree species are listed under their scientific names only.